Strategic Leadership in Responsive Web Design

A tech leader's guide to achieving business excellence by adopting responsive web design

Harley Ferguson

Strategic Leadership in Responsive Web Design

Group Product Manager: Kaustubh Manglurkar
Publishing Product Manager: Vaideeshwari Muralikrishnan
Book Project Manager: Aishwarya Mohan
Senior Editor: Debolina Acharyya
Technical Editor: Simran Ali
Copy Editor: Safis Editing
Indexer: Rekha Nair
Production Designer: Ponraj Dhandapani
DevRel Marketing Coordinators: Anamika Singh and Nivedita Pandey

First published: September 2024

Production reference: 1020824

Published by Packt Publishing Ltd.
Grosvenor House
11 St Paul's Square
Birmingham
B3 1RB, UK

ISBN 978-1-83508-078-8

www.packtpub.com

Contributors

About the author

Harley Ferguson is a software development agency co-founder, CEO, developer, author, and speaker. Harley started his journey of coding at the age of 9 and never looked back. Having built his way from an intern developer to a technical lead in a matter of years, Harley moved on to start Origen Software, a bespoke development agency based out of Cape Town, South Africa. Harley works closely with startups, helping them build their businesses and products through software. Harley is also an investor and advisor to several startups. Beyond writing code, Harley also shares his developer insights and advice publicly, which has resulted in him being one of the largest developer influencers in the world.

About the reviewers

Darian Rory Leonard is a storyteller, a truth seeker, and occasionally a frontend software developer, working to increase the priority of implementing good UI/UX. Starting his career in accounting before pivoting into software development, he's a walking testament to the reality that the barrier to entry for software development has never been lower and there's never been a better time to be a developer. Always striving to leave a project better than he found it, he advocates for developers to write simple code that solves the problem before code demanded by any ego. He had a short-lived start in Serendipity, a pitstop in C#, and settled comfortably in the frontend suburbia of React, with a holiday home in BDE.

Ricardo Zea is a seasoned senior web designer with extensive experience in all aspects of web design: ideation, sketching, wireframing, designing, prototyping, hand-off to development, production-ready HTML and CSS, and maintenance. He has worked with Packt as an author creating the books *Mastering Responsive Web Design* and *Web Developers Reference Guide*, and as a technical reviewer on this book and *Mastering Sass*. He also wrote, for Smashing Magazine, an article called *An Ultimate Guide To CSS Pseudo Classes And Pseudo Elements*. Furthermore, he has also worked for several large companies including LexisNexis, Nestlé, and Willis Towers Watson. Since 2016, Ricardo has also been a Web Design expert mentor on Codementor with a 5-star rating.

Table of Contents

3

Design Systems and Responsiveness   53

4

Content and Media Strategy for Varied Screens   69

5

Optimization and Performance — a Business Imperative   91

6

Strategic Breakpoints and Adaptability   115

7

User Navigation and Conversion Pathways   131

8

The Business Case for Web Accessibility   149

9

SEO Considerations in Responsive Design   175

10

Leading and Guiding Design Teams   191

11

Quality Assurance and ROI   217

12

Vendor and Platform Decisions in RWD   235

13

Engaging Users Across Touchpoints   249

14

Future-Proofing Responsive Strategies   265

15

Building a Responsive Culture 283

Preface

Welcome to a comprehensive exploration of **responsive web design** (RWD), a critical aspect of modern web development that ensures websites are accessible and functional across a variety of devices. As the digital landscape continues to evolve, the ability to create web interfaces that adapt dynamically to different screen sizes and orientations is not just an advantage; it's also a necessity. This book delves into the strategic implementation of RWD, providing tech leaders with the insights needed to harness its full potential.

RWD is about more than just making a website look good on a mobile phone or tablet; it's also about creating seamless experiences that enhance user satisfaction and drive business outcomes. Through a series of detailed chapters, this book covers the fundamentals of RWD, from planning and design to deployment and optimization. Each section is crafted to equip you with the skills and knowledge to not only implement responsive strategies but also to lead teams in the development of websites that stand out in versatility and user engagement. Whether you're looking to refine your approach or build a responsive culture within your team, this book offers the guidance necessary to navigate the responsive web landscape effectively.

Who this book is for

This book is designed for senior tech leaders and decision-makers who are involved in guiding their teams through the complexities of RWD. Ideally, you should have a solid background in web development or project management within the tech industry. This includes individuals such as **Chief Technical Officers (CTOs)**, lead developers, and project managers who are looking to deepen their understanding of RWD strategies and implement them effectively to enhance user experience and meet business goals. The content is tailored to help these professionals leverage RWD as a strategic tool in their organizations, ensuring they are well-equipped to make informed decisions and drive successful projects.

What this book covers

Chapter 1, Responsive Web Design – a Strategic Overview, provides an overarching view of how web design has evolved and how businesses can leverage RWD for strategic benefits. You will gain a deeper understanding of the pivotal role RWD plays in aligning with business objectives.

Chapter 2, Mobile-First Strategy in a Multi-Device World, delves into why prioritizing mobile design is paramount. You will appreciate the **return on investment (ROI)** from a mobile-centric approach and its ripple effects.

Chapter 3, Design Systems and Responsiveness, emphasizes the synergy between design systems and RWD, providing leaders with insights into ensuring scalability and consistency.

Chapter 4, Content and Media Strategy for Varied Screens, dives into how to make content resonate across devices, ensuring users are engaged and businesses achieve conversion goals.

Chapter 5, Optimization and Performance – a Business Imperative, explores the intrinsic link between web performance, user retention, and overall business success.

Chapter 6, Strategic Breakpoints and Adaptability, dives into how to strategically select breakpoints, ensuring seamless transitions and optimal user experiences.

Chapter 7, User Navigation and Conversion Pathways, explores the tenets of intuitive navigation and how it dovetails with conversion strategies across devices.

Chapter 8, The Business Case for Web Accessibility, introduces the business value of accessible design and how it overlaps with responsive strategies.

Chapter 9, SEO Considerations in Responsive Design, unpacks how responsive design can affect SEO rankings, providing a roadmap to ensure that sites are discoverable and performant.

Chapter 10, Leading and Guiding Design Teams, explores gestures such as swipe and pinch and their importance in modern web applications.

Chapter 11, Quality Assurance and ROI, seeks to equip you with strategies to lead design teams effectively, fostering innovation and aligning with business objectives.

Chapter 12, Vendor and Platform Decisions in RWD, navigates the myriad of vendor and platform options available, ensuring your decisions align with both immediate needs and future scalability.

Chapter 13, Engaging Users Across Touchpoints, delves into strategies to captivate users, regardless of their device or platform.

Chapter 14, Future-Proofing Responsive Strategies, equips you with the knowledge to stay ahead, ensuring your responsive strategies remain relevant, effective, and adaptable to emerging trends.

Chapter 15, Building a Responsive Culture, explores how to permeate this philosophy throughout your organization, making responsiveness a part of its very culture.

To get the most out of this book

To get the most out of this book, it's essential to have a foundational understanding of web development principles and technologies. You should be familiar with HTML, CSS, and JavaScript, as these are the core tools used to implement RWD. Basic knowledge of web development frameworks and pre-processors can also be beneficial, as they are often discussed in the context of streamlining and enhancing responsive workflows.

Additionally, some experience with **User Interface (UI)** and **User Experience (UX)** design will help in understanding the discussions about creating optimal user interactions across different devices. Understanding the basics of SEO and how it integrates with web design is also assumed, as it is crucial for the successful deployment of any web project.

This book assumes that you are comfortable with these concepts and focuses on applying them within the framework of RWD, enabling you to elevate your projects and lead your teams effectively in today's multi-device world.

Conventions used

`Code in text`: Indicates code words in text, database table names, folder names, filenames, file extensions, pathnames, dummy URLs, user input, and Twitter handles. Here is an example: "The `srcset` HTML attribute allows developers to specify multiple image sources for different screen sizes."

Bold: Indicates a new term, an important word, or words that you see onscreen. For instance, words in menus or dialog boxes appear in **bold**. Here is an example: "Product details, reviews, and the all-important **Add to Cart** button remain accessible and prominent."

> **Tips or important notes**
> Appear like this.

Get in touch

Feedback from our readers is always welcome.

General feedback: If you have questions about any aspect of this book, email us at `customercare@packtpub.com` and mention the book title in the subject of your message.

Errata: Although we have taken every care to ensure the accuracy of our content, mistakes do happen. If you have found a mistake in this book, we would be grateful if you would report this to us. Please visit `www.packtpub.com/support/errata` and fill in the form.

Piracy: If you come across any illegal copies of our works in any form on the internet, we would be grateful if you would provide us with the location address or website name. Please contact us at `copyright@packt.com` with a link to the material.

If you are interested in becoming an author: If there is a topic that you have expertise in and you are interested in either writing or contributing to a book, please visit `authors.packtpub.com`.

Share Your Thoughts

Once you've read *Strategic Leadership in Responsive Web Design*, we'd love to hear your thoughts! Scan the QR code below to go straight to the Amazon review page for this book and share your feedback.

https://packt.link/r/1835080782

Your review is important to us and the tech community and will help us make sure we're delivering excellent quality content.

Download a free PDF copy of this book

Thanks for purchasing this book!

Do you like to read on the go but are unable to carry your print books everywhere?

Is your eBook purchase not compatible with the device of your choice?

Don't worry, now with every Packt book you get a DRM-free PDF version of that book at no cost.

Read anywhere, any place, on any device. Search, copy, and paste code from your favorite technical books directly into your application.

The perks don't stop there, you can get exclusive access to discounts, newsletters, and great free content in your inbox daily

Follow these simple steps to get the benefits:

1. Scan the QR code or visit the link below

https://packt.link/free-ebook/9781835080788

2. Submit your proof of purchase
3. That's it! We'll send your free PDF and other benefits to your email directly

1

Responsive Web Design — a Strategic Overview

Welcome to the first chapter, where we embark on a journey to explore the pivotal world of **responsive web design (RWD)**. This chapter isn't just about understanding the technicalities; it's about recognizing RWD as a strategic asset in today's digital landscape.

In the coming sections, we will delve deep into the evolution of RWD, understanding how it has transformed from a mere design approach to a crucial business strategy. You'll not only learn the fundamental principles that underpin RWD but also explore its implications for **user experience (UX)**, **search engine optimization (SEO)**, and overall business impact.

Practically, you'll be equipped with insights to assess and strategize RWD for your projects or organization. This includes learning how to evaluate your current digital assets, understanding the steps to transition to a responsive framework, and identifying **key performance indicators (KPIs)** to measure the success of your RWD initiatives.

But why is this important? In a world where digital presence is no longer a luxury but a necessity, RWD stands as a cornerstone for success. It's not just about making websites look good on different devices; it's about ensuring that your digital offerings are accessible, efficient, and engaging, regardless of how your audience chooses to access them. This chapter will lay the foundation for you to build a web presence that is not just responsive in design but also in meeting the dynamic needs of your audience and business goals.

As we progress, you'll gain both a strategic and practical understanding of RWD. This isn't just about learning; it's about applying these insights in real-world scenarios. Whether you're a business leader, a developer, or a designer, this chapter will provide you with the tools and knowledge to leverage RWD as a key driver for digital success. Let's begin this journey toward creating a more adaptable, accessible, and engaging digital world.

The topics covered in this chapter are the following:

- The evolution of web usability
- Business benefits of implementing RWD
- Aligning RWD with business objectives
- Challenges in implementing RWD

The evolution of web usability

When reminiscing about the 90s, we often conjure images of grunge music, flannel shirts, and the growing prominence of MTV. But it was also an era that bore witness to the birth of the consumer internet — a world filled with rudimentary web pages and slow, laborious dial-up connections. These initial websites, often crafted by enthusiastic hobbyists or forward-thinking businesses, were little more than static digital brochures. Typography was elementary, images were pixelated, and user interactivity was almost non-existent.

The late 90s saw the beginning of businesses scrambling to create their online presence. As they all started venturing into the online world, websites became more complex and interactive. Companies began implementing technologies such as *Flash*, which was used to create vibrant animations, embedded videos, and a level of interactivity previously unseen. Businesses were hoping that they could impress users with their intricate designs and dazzling elements.

However, there was a flip side to this coin. Many users found themselves lost in these new websites, struggling to navigate their complexity. Bounce rates soared, and user frustration was palpable. It was clear that while digital aesthetics were advancing, the foundational principles of UX were being sidelined. A balance needed to be struck — a balance where aesthetics met usability, where form harmonized with function.

The dawn of the 21st century saw a shift in the mindset of web developers and designers worldwide. No longer was the creation of a website just a technical endeavor; it became a holistic process that married the technical with the psychological. Pioneers in the field began to argue that understanding user behavior was paramount. They championed the cause of user-centric design, where the needs and behaviors of the end users shaped the website's architecture and flow.

During this era, the term *usability* graduated from being a mere industry buzzword to a guiding principle. Usability encapsulates the idea that a website should be intuitive, be easily navigated, and deliver value to its users without unnecessary complexities. As businesses recognized the correlation between usability and customer satisfaction, investments in this field skyrocketed.

The late 90s and early 2000s saw the rise of usability labs. These specialized setups, often found within tech companies or as standalone research entities, were designed to study how real users interacted with digital products. Armed with eye-tracking software, heatmaps, and behavioral analysis tools, these labs played a pivotal role in shaping the next generation of web design. By observing user interactions and analyzing points of friction, designers could refine their creations, ensuring that websites were not just beautiful but also functional and user-friendly.

Let's explore the rise of mobile devices.

The mobile revolution and the usability paradigm shift

The mid-2000s marked a significant turning point in the world of digital design. The release of smartphones, led by Apple and its iPhone, triggered a seismic shift in how users accessed and interacted with the web. Users were no longer confined to only accessing the web through their desktops. The internet became something that millions of users could access from a device in their pocket. However, this mobile revolution brought with it an array of new usability challenges that designers had not previously faced.

The diversity of screen sizes introduced a lot of complications. Websites designed for the expansive real estate of desktop monitors now had to function seamlessly on smaller, touch-based interfaces. The pinch, swipe, and tap gestures that seem so simple today were a new way of interacting with websites that hadn't previously been seen. Desktops were only making use of cursor-based navigation.

Businesses were faced with a dilemma: should they maintain one version of their website for desktop users and create an entirely separate mobile version? Many early adopters of mobile web design opted for this, resulting in the *m-dot* websites, where a mobile-specific version of a site was hosted on a subdomain.

As devices continued to be rolled out in various sizes, maintaining multiple versions of a website proved pretty much impossible. This challenge birthed the concept of RWD. Instead of creating different versions for each screen size, RWD allowed websites to adapt to different screen sizes, ensuring a consistent and usable experience across devices.

Ethan Marcotte, credited with coining the term *responsive web design* in his 2010 article (`https://alistapart.com/article/responsive-web-design/`), described it as a trio of techniques: a fluid grid, flexible images, and media queries. This approach was revolutionary, offering a more sustainable and user-centric solution to the multi-device world's challenges. No longer would multiple variants of the same website need to be built and maintained to cater to varying device sizes.

Businesses quickly realized that a mobile-friendly web presence was no longer a luxury — it was a necessity. User expectations were evolving, and patience for clunky, non-responsive sites was waning. A poor mobile experience could mean lost customers and diminished brand perception. In this era of web accessibility through mobile phones, good usability was not just about user satisfaction; it directly impacted the bottom line.

E-commerce sites, in particular, felt the pressure. The rise of mobile shopping meant that product displays, checkout processes, and payment gateways had to be rethought for smaller screens and touch interactions.

Things only got more complex as smartphones hit the market.

The rise of devices and the challenge of adaptability

Post-2010, the mobile revolution was in full swing. But it wasn't just about smartphones anymore. Tablets began to take center stage, blurring the lines between phones and full-fledged computers. Companies such as Apple, Samsung, and Microsoft launched devices that ranged vastly in size, from compact 5-inch screens to larger 12-inch displays.

This diversification meant that web designers weren't simply dealing with *mobile versus desktop* anymore. The new spectrum of device sizes introduced the need for a website to cater to a further array of screen sizes and resolutions.

Tablets presented a unique challenge. They offered more screen real estate than phones, allowing for richer designs and content layouts. But they also retained touch-based interaction, requiring designers to find a balance between the compactness of mobile and the expansiveness of desktop interfaces.

Some industries, notably e-magazines and digital publishing, flourished on tablets as they could mimic the familiar format of printed material while leveraging interactive digital elements. This presented opportunities for innovative design but also demanded a new level of adaptability.

While mobile and tablet browsing soared, desktops retained their importance, especially for tasks deemed productive. E-commerce sites saw a pattern where users might discover products on mobile but switch to desktops to finalize purchases. This cross-device journey emphasized the importance of a consistent, responsive experience, ensuring users could pick up where they left off, regardless of the device.

Different devices often mean different user behaviors. A smartphone user on the go might want quick access to information, while a tablet user lounging at home might be more inclined toward deeper dives into content. Recognizing these patterns was crucial. Websites had to be more than just visually adaptable; they needed to cater to the diverse intentions and expectations of users across devices.

This era of device proliferation underscored a key lesson: adaptability was not just about fitting content on varied screens but truly understanding and catering to the nuanced behaviors of users across devices. As tech leaders, recognizing this intricacy would be pivotal in shaping the next wave of web design strategies.

Now, let's break down what modern web usability is.

Modern web usability – a multifaceted approach

In the realm of RWD, one of the foundational truths that tech leaders have come to realize is the diversity of the modern web user. Beyond devices, today's internet dweller juggles various identities — a reader, a shopper, a researcher, a creator, and often, a multitasker. The challenge? Crafting websites that can intuitively serve all these needs.

Touchscreens, voice assistants, **augmented reality (AR)** integrations – the modern web isn't just visual. Interactivity has taken center stage. For businesses, this translates to not only creating aesthetically pleasing sites but also integrating interactive elements that are intuitive and enhance user engagement. Dynamic elements such as chatbots, scroll animations, or real-time experience customization features are no longer add-ons but expectations.

A truly modern web experience is inclusive. Tech leaders now recognize the profound business and ethical imperatives of web accessibility. Sites are now designed keeping in mind users with varied abilities, ensuring a seamless experience for everyone, whether they're navigating through a screen reader, voice commands, or other assistive technologies.

The rise of analytics tools and AI-powered insights has transformed web usability from an art to a precise science. Modern sites are not just built; they're iterated upon. Real-time user behavior data, A/B testing outcomes, and heatmaps guide the refining process, ensuring that design choices align closely with user preferences and business objectives.

That's how we do things now, but how will things change in time?

The future of web usability and the importance of continued evolution

It's an accepted reality: the only constant in the digital world is change. The swift pace at which technology evolves necessitates a proactive approach from tech leaders, ensuring that their strategies not only address present challenges but also anticipate future shifts. As we stand on the verge of a new era of web design, the horizon is rife with innovations waiting to redefine how users interact with the web.

Multiple areas will take web usability to the next level in the future.

Quantum computing, with its revolutionary processing power and capabilities, promises to redefine web usability. Tech leaders must grapple with the implications of near-instantaneous data processing, AI-driven design optimizations, and the potential for fully immersive web experiences that blur the line between virtual and reality.

While AR and **virtual reality (VR)** technologies have made inroads in gaming and specific industries, their potential in general web design remains largely untapped. The idea of a website transforming into an interactive 3D space or overlaying information on the real world presents unprecedented engagement opportunities. Adapting to this will require leaders to rethink design from the ground up.

Research suggests that by 2030, nearly 20% of web interactions could involve some form of AR, seamlessly merging the physical and digital worlds.

The future isn't just about flashy tech innovations. As societal awareness around sustainability grows, the push for green and ethical web design will become pronounced. This means optimizing websites to consume less energy, ensuring ethical data-handling practices, and creating designs that prioritize users' mental well-being.

One clear message echoes through the annals of web design's history: adapt or risk obsolescence. For tech leaders, this underscores the importance of continuous learning, staying abreast with emerging trends, and fostering a culture of adaptability within teams. The journey of RWD isn't a destination but a continuous evolution, one that promises to keep the digital realm dynamic and user-centric.

Now that we've explored the history of RWD, let's talk about how to use it to benefit your business.

Business benefits of implementing RWD

The dawn of the digital age promised many things: a connected world, instant communication, and a wealth of information at our fingertips. As we've navigated this evolving landscape, one truth has become abundantly clear — technology must serve the user, not the other way around. This principle has reshaped industries, toppled giants, and given rise to new champions. Nowhere is this more evident than in the realm of web design, where the user-centric paradigm has not just been a trend, but a transformative force.

The ethos of RWD is deeply rooted in this user-centric philosophy. It's not about mere aesthetics or following the latest design trend; it's about creating a seamless, intuitive experience for users regardless of the device they're using. Think about it: what's the use of the most visually stunning website if half your audience can't navigate it properly on their smartphones or tablets?

In the early days of the internet, websites were designed primarily for desktop users. Monitors had relatively standardized sizes, and designers could make reasonable assumptions about how a site would be viewed. But the rapid proliferation of devices — from mobile phones with tiny screens to large tablets and everything in between — threw a wrench into this model. Websites that looked good and functioned beautifully on a desktop could be virtually unusable on a mobile device.

Businesses faced a choice: adapt to implement RWD or risk alienating a significant portion of their audience. Those who recognized the importance of the UX and prioritized it flourished, while those who clung to the old ways found themselves struggling to maintain relevance. This is the crux of the user-centric paradigm — it's a business survival tactic as much as it is a design philosophy.

Let's consider a practical example. Imagine you're a busy professional trying to make a quick online purchase during your commute. You pull up a website on your phone, only to find the text is too small to read, the buttons are too close together to tap accurately, and every attempt to navigate sends you to an unintended page. Frustrated, you give up and decide to buy from a competitor whose site is more mobile-friendly. The first business lost a sale, not because of its products or prices but because its website failed to prioritize the UX.

This isn't just about sales. It's about trust, credibility, and building lasting relationships with customers. A responsive website signals to users that a business understands and cares about their needs. It suggests professionalism, attention to detail, and a commitment to quality. In today's digital age, where users often interact with businesses first and foremost through their online presence, this can make or break a brand's reputation.

But it's not just businesses that benefit. When websites are designed with users at the forefront, it empowers individuals. It means easier access to information, smoother transactions, and a generally more pleasant online experience. In essence, a responsive, user-centric design approach democratizes the web, making it more accessible and useful for everyone.

As tech leaders, understanding the strategic importance of the user-centric paradigm is crucial. It's not just a matter of staying current with design trends; it's about recognizing the profound impact of the UX on business success. As we delve deeper into the benefits of a responsive approach in the coming pages, always remember that at its core, RWD is about people. It's about serving their needs, respecting their time, and delivering value at every touchpoint.

This perspective shift — from business-centric to user-centric — has set the stage for the numerous benefits of RWD that we will explore next: benefits that not only enhance the UX but also drive tangible business outcomes, from increased engagement and sales to enhanced brand trust and loyalty.

It's time to talk about increasing engagement and reducing bounce rates.

Increased engagement and reduced bounce rate

Navigating the intricate channels of the digital age often seems analogous to steering a ship through treacherous waters. The myriad of challenges — changing user preferences, evolving technology, and a plethora of competition — demands a robust strategy. Amid this tumultuous sea, there's a beacon that businesses constantly strive toward — user engagement. And RWD, with its user-centric philosophy, proves to be an invaluable compass guiding toward this very beacon.

Engagement is more than just an industry buzzword. It's the lifeblood of the digital realm. A website that lacks engagement is a website that won't last long. When users are actively engaged, they interact with content, share it, comment on it, and return for more. Every click, every share, every minute spent on a site is a testament to its ability to capture attention and provide value. In an age of information overload, this is no small feat.

On the flip side, there's the dreaded *bounce rate* — *a* metric that signifies visitors leaving a site after viewing just one page. It's the digital equivalent of walking into a store, glancing around, and walking right back out. High bounce rates are often symptomatic of deeper issues: poor design, irrelevant content, or a non-intuitive UX.

So, how does responsive design fit into this narrative? Let's begin with a simple premise. The more user-friendly a site is across various devices, the more likely users are to engage with its content. It's not rocket science. If a site seamlessly adjusts to fit the screen of a smartphone or tablet, providing clear text and easily clickable links, users are bound to spend more time exploring.

The connection between responsive design and engagement is bolstered by numerous studies. Research consistently shows that websites optimized for mobile viewing have lower bounce rates and longer average session durations compared to their non-responsive counterparts. It's a clear indication that users appreciate and reward the effort businesses put into crafting a seamless browsing experience.

Beyond just metrics, there's a subtler, yet equally significant advantage. When a site delivers a consistent UX across devices, it builds trust. Users come to rely on the site's functionality and design, knowing that whether they access it from their desktop at work, tablet on the couch, or phone in a coffee shop, they'll be met with the same level of excellence.

For businesses aiming to establish a strong digital presence, these factors can't be ignored. In the upcoming pages, we'll dive into the nuanced relationship between responsive design and other KPIs, unraveling the various benefits of a user-first approach. As the boundaries between devices blur and the digital ecosystem continues to evolve, one thing remains clear: those that prioritize the UX will always stand a step ahead in the engagement game.

That leads me to my next point: your brand is your business.

Enhanced brand reputation and trust

In the vast tapestry of the digital ecosystem, brand reputation operates as a cornerstone of success. Every online interaction, be it on a website, social platform, or via email, not only offers a service or conveys a message but also shapes the perception of a brand. Brands that can consistently provide quality experiences inevitably carve out a place in the collective consciousness of their audience. And when RWD enters the narrative, it amplifies this process, intertwining functionality with brand trust.

Reflect for a moment on personal experiences. How often have you equated a seamless, intuitive website experience with the quality of the product or service the brand offers? Subconsciously, users form associations: *If a company invests effort into making its website so user-friendly, surely its products and services are of similar quality.*

This psychological connection, while understated and somewhat simple, is profoundly influential. In the age of information, where reviews, feedback, and UXs are just a click away, businesses cannot afford to overlook the impact of digital touchpoints on brand reputation.

Responsive design is not just about making things look good; it's about feeling good too. When users can transition from their desktop to their mobile phone without missing a beat, experiencing the same high-quality content tailored to their device, it fosters a sense of familiarity. Over time, this familiarity translates into trust. Just as people trust familiar faces in a crowd, they trust brands that consistently deliver on their promise of quality, irrespective of the medium.

Picture a person walking into a luxury store that sells high-end watches. They are there to buy the product, but they are also there for the experience that comes along with the purchase. Using RWD and branding to build an impactful UX is the digital equivalent of this.

A responsive design communicates more than just adaptability. It sends a message that a brand is updated, in step with technology, and attentive to its users' needs. It indicates foresight — recognizing that users today are as likely, if not more, to engage with content on their phones as they are on their PCs.

But the benefits of responsive design aren't limited to luxurious or tech-savvy brands. Regardless of industry or target demographic, ensuring a consistent, adaptive digital experience directly feeds into a positive brand reputation. It communicates care, commitment, and professionalism.

We'll delve deeper into real-world scenarios in subsequent pages, exploring the varied dimensions of brand trust in the digital realm. Through case studies and data-backed insights, we'll witness the transformative potential of responsive design, not just as a tech solution but as a strategic brand enhancer. As the chapters unfold, one point will be unequivocally clear: in the intricate dance of digital perception, responsive design leads the way, choreographing moves that resonate with the rhythm of user trust.

This leads us to think about conversion.

Optimized conversion pathways

Conversion. At its core, this simple word encapsulates the goals of countless businesses in the digital realm. Whether it signifies a sale, a sign-up, a subscription, or any other desired action, conversion represents the culmination of efforts to guide a user from casual browsing to commitment. Yet, in the labyrinth of digital interfaces, ensuring a smooth and direct conversion pathway is no small feat. Enter RWD — a tool not only for aesthetic adaptability but also for honing the journey toward conversion.

The origin of conversion optimization lies in understanding user behavior. How do users navigate a website? Where do they hesitate, and why? What motivates them to take action? In piecing together this puzzle, it becomes clear that a singular, static design approach cannot account for the multitude of devices and screen sizes users employ. RWD, in its essence, adapts to the user, and in doing so, simplifies their pathway to conversion.

Consider the scenario of an online retailer. A potential customer might discover a product while browsing on a desktop during a lunch break, only to revisit and make the purchase on a mobile device during the evening commute. A responsive website ensures that this user doesn't face any hindrances or inconsistencies when switching devices. Product details, reviews, and the all-important **Add to Cart** button remain accessible and prominent. The transaction process remains streamlined and intuitive. The result? A higher likelihood of conversion, all thanks to a design that adapts to the user's context.

This adaptability is more than just about screen size; it's about creating an environment conducive to action. When users feel at ease, when they aren't pinching, zooming, or squinting to navigate, they're more focused on content. When content — from product listings to blog posts — is the center of attention, users are better positioned to respond to **calls to action (CTAs)**.

Moreover, an optimized conversion pathway reaps benefits beyond the immediate sale or sign-up. It contributes to a more favorable UX, which, in turn, boosts return visits and referrals. Think of it as laying down a golden pathway, where every step the user takes feels natural, intuitive, and inviting.

However, it's crucial to recognize that while RWD provides the foundation for optimized conversion pathways, continuous analysis and iteration are key. User behaviors evolve, technologies advance, and businesses must remain agile, always fine-tuning their approach to guide users effectively.

How can we achieve this? Let's find out.

Futureproofing your digital presence

The digital realm is in constant flux. This is a given. Just as businesses adjust to one technological advancement, another emerges on the horizon, promising even more revolutionary changes. This ceaseless evolution poses a challenge for businesses: How can one build a digital presence that not only thrives today but remains relevant and functional in the years to come? This is where the concept of futureproofing enters the equation, and RWD stands at its core.

At a glance, the term *future-proof* suggests an unchanging state, but in the digital context, it's quite the opposite. Futureproofing entails the ability to adapt, evolve, and embrace whatever the future brings, ensuring that investments made today yield value for years to come. With RWD, businesses lay the groundwork for such adaptability.

First and foremost, the very philosophy of RWD is rooted in flexibility. By removing fixed widths and rigid layouts, responsive designs flow and reshape based on the user's device. This fluidity ensures that as new devices emerge — be it larger desktop monitors, foldable smartphones, or even wearables — a responsive website can accommodate them without the need for drastic overhauls.

Beyond mere screen sizes, RWD also promotes a modular approach to web content and functionality. By thinking in terms of reusable components rather than fixed templates, businesses can swiftly integrate new features or content types as they become relevant. This modular mindset empowers businesses to ride the wave of technological advancements, integrating innovations such as AR modules or voice-command interfaces as they mature and become mainstream.

Yet, futureproofing is not solely a technological endeavor; it also demands a forward-thinking approach to UX. User behaviors and expectations evolve alongside technology. What's considered intuitive and user-friendly today might feel cumbersome or dated in just a few years. Here, the principles of RWD come into play, emphasizing user-centric design, clear hierarchies, and a focus on core functionalities. These foundational elements serve as a stable base upon which evolving user trends can be addressed and integrated.

Take, for instance, the rise of voice search or command. A few years ago, this was a niche feature. Today, with the proliferation of smart speakers and voice-activated assistants, it's becoming a primary mode of web interaction for many. A future-proofed responsive site would not only ensure compatibility with these voice platforms but would also optimize content to be easily discoverable and consumable via voice.

However, it's essential to strike a balance. In the pursuit of futureproofing, businesses shouldn't become early adopters blindly. Not every emerging technology will become mainstream, and not every trend will have lasting power. The key is to stay informed, discern the potential long-term impact of new developments, and integrate them judiciously into the digital strategy.

How can we achieve all of this while not going over budget? Let's see.

Cost efficiency and maintenance

In a world of accelerating digital transformation, financial prudence is a virtue that businesses cannot afford to ignore. Investments in technology and digital infrastructure, while crucial, must be approached with an eye toward long-term value and sustainability. This is where RWD shines. Beyond the immediate and evident advantages in UX, RWD offers compelling benefits in terms of cost efficiency and streamlined maintenance.

At the outset, the savings from adopting RWD might not be immediately palpable. There are costs associated with redesigning an existing site or building a new responsive platform from the ground up. However, when viewed through a longer-term lens, the value proposition becomes strikingly clear.

Traditional approaches to web design often involved creating separate sites for different devices. This meant businesses had to manage and maintain multiple code bases, each tailored for desktops, tablets, or mobiles. With RWD, the game changes. A single code base can fluidly adapt across devices. This unified approach translates into easier maintenance. Updates, changes, or fixes need to be implemented just once, not multiple times across different versions.

The tech landscape is unpredictable. New devices with varied screen sizes and capabilities can emerge at any moment, rendering obsolete sites designed for yesterday's devices. In such a scenario, constant redesigns and tweaks become a financial drain. RWD offers an adaptive foundation. As devices evolve, a responsive site naturally adjusts, obviating the need for frequent, costly overhauls.

Performance optimization is a critical concern for businesses. Slow-loading sites can deter users and hurt conversion rates and search engine rankings. With separate sites for different devices, businesses often find themselves optimizing performance multiple times over. RWD streamlines this. Since there's a single site serving all devices, performance optimizations become centralized, leading to time and cost savings.

Growth is the aim of any business, and with growth comes inevitable website updates, whether they're new sections, features, or functionalities. With a responsive setup, scaling becomes more straightforward. The modular approach inherent to RWD means new elements can be integrated seamlessly without disrupting the existing UX or necessitating substantial code reworks.

Uniformity breeds predictability. With just one site to monitor and maintain, IT and design teams can establish regular, predictable maintenance cycles. This regularity ensures that potential issues are identified and addressed promptly, leading to reduced downtimes and ensuring a consistent UX.

Yet, it's important to approach the cost aspect of RWD with nuance. While there are undeniable savings and efficiencies, businesses must also budget for periodic reviews, UX testing, and staying abreast of emerging RWD trends and best practices. These investments, while incurring some costs, are essential to ensure the long-term viability and effectiveness of a responsive site. Implementing RWD cannot be seen as a once-off investment but rather a continuous exercise to help ensure longevity.

In conclusion, RWD, when approached strategically, can be a game-changer in maximizing your **return on investment** (**ROI**). It reduces redundancies, fosters efficiency, and, most importantly, positions businesses favorably in a constantly evolving digital ecosystem.

Now, how can we ensure traffic? Let's find out.

SEO and visibility in the digital marketplace

In today's digital age, a strong online presence isn't just about having a well-designed website or a robust product offering. It's about being visible and accessible where your audience is searching. Enter the realm of SEO — the art and science of enhancing a website's visibility in search engine results. In the context of RWD, SEO takes on a nuanced significance, forging a synergy that tech leaders need to harness for true digital success.

Google, the titan of search engines, made a pivotal shift between 2016 to 2018 by introducing mobile-first indexing. This meant that the mobile version of a website became the primary point of evaluation for ranking. With the majority of users accessing the web via mobile devices, this shift was both intuitive and inevitable. RWD, with its inherent mobile adaptability, dovetails perfectly into this paradigm, ensuring websites are naturally primed for mobile-first indexing criteria.

Search engines prioritize UX. Slow-loading pages, which are a deterrent for users, consequently affect search rankings. RWD optimizes images and elements to suit the accessing device, ensuring faster load times. When a website is responsive, it inherently boosts page loading speeds, especially on mobile devices, contributing positively to SEO.

In the pre-RWD era, businesses often created separate mobile and desktop versions of their websites. This dual setup inadvertently led to content duplication, which search engines frowned upon. With RWD, the content remains consistent across devices, eliminating issues of duplication and the associated SEO pitfalls.

The bounce rate represents the percentage of visitors who navigate away from a site after viewing just one page. High bounce rates can be indicative of poor UX, which can negatively impact SEO. RWD ensures a seamless experience across devices, potentially reducing bounce rates and signaling to search engines that the site is user-friendly.

For businesses targeting local markets, RWD offers a distinct advantage. Mobile users often search for local information on the go. A responsive site, which provides an optimized mobile experience, is more likely to engage these users, bolstering local SEO rankings.

But why does all this matter from a business perspective? In a cluttered digital marketplace, visibility is currency. The higher a site ranks on search engines, the more organic traffic it attracts. And organic traffic, often representing users genuinely interested in a product or service, tends to have higher conversion rates.

Furthermore, in a world dominated by **pay-per-click** (**PPC**) advertising, organic rankings provide a cost-effective way to reach audiences. While SEO does entail consistent effort and optimization, its returns, especially when combined with RWD, can be substantial.

To distill it down, in the interplay between RWD and SEO, tech leaders find a potent combination to amplify their brand's digital footprint. By understanding and leveraging this synergy, businesses can not only enhance their digital visibility but also establish a foundation for sustained online growth and engagement in the ever-competitive digital marketplace.

Next, we'll discuss how to ensure your business objectives align with RWD.

Aligning RWD with business objectives

The term *digital transformation* has been a buzzword in the corporate landscape for a while now, often evoking images of radical shifts in business processes, leveraging advanced technologies, and a general move toward a more digitized business environment. At the heart of this transformative journey, however, lies a fundamental shift in how businesses interact with their consumers in the digital space. As these digital interactions become the primary mode of engagement, the importance of RWD in this transformational process cannot be overstated.

Let's first establish what digital transformation truly entails. It's not just about adopting new technologies or migrating to the cloud; it's about fundamentally rethinking how a business delivers value to its customers in a digital-first world. This shift demands agility, flexibility, and an unwavering focus on UX. After all, in a world where consumers have more choices than ever and can switch brands with a simple click, ensuring a smooth, consistent digital experience becomes a competitive advantage.

Enter RWD.

In the early days of the internet, websites were designed for a single, standard desktop screen size. As devices proliferated, from smartphones to tablets, wearables to smart TVs, the digital landscape fragmented. Businesses found themselves in a conundrum. Creating a separate website for each device type was impractical, expensive, and a maintenance nightmare. Yet, providing a subpar UX on any device was not an option in the age of the discerning digital consumer.

RWD emerged as a strategic solution to this problem, but its implications were far more profound than just flexible layouts. It became a cornerstone of the broader digital transformation strategy for forward-thinking businesses. By adopting a responsive approach, companies were signaling to their users that they understood the evolving digital landscape. They were committed to providing a consistent, high-quality experience irrespective of the device used.

Furthermore, RWD's adaptive nature aligns seamlessly with the ethos of digital transformation: agility. Just as businesses must pivot and adapt to market changes, technological advancements, or competitive pressures, so too must their digital platforms. With RWD, businesses can ensure that their online presence is not just adaptable for today's devices but also future-ready for whatever innovations tomorrow might bring.

Another critical aspect of digital transformation is data-driven decision-making. With responsive design, businesses can gain richer insights into user behavior across devices. How does mobile engagement differ from desktop? At what screen size do users most commonly drop off? These insights, powered by RWD, can inform broader strategic decisions, from marketing campaigns to product launches.

In conclusion, while RWD started as a technical solution to a design problem, its role in the broader digital transformation narrative is pivotal. As businesses strive to redefine their value proposition in a digital-first world, ensuring a seamless, consistent, and adaptive digital experience through RWD is not just an IT decision — it's a strategic imperative.

That's why you need buy-in from all crucial decision-makers.

Identifying key business stakeholders

Every project within an organization, no matter how technical or design-oriented, operates with varied interests, goals, and challenges. The story is no different when considering the rollout or redefinition of an RWD strategy. Beyond the web developers and UI/UX designers, numerous stakeholders play critical roles in ensuring the success of such a venture. Understanding and identifying these stakeholders is fundamental to aligning RWD with business objectives.

The term *stakeholder* might evoke the image of a boardroom filled with executives, but in the context of RWD, it's far more encompassing. Stakeholders are those individuals or groups who have a vested interest in the outcome of the RWD project. Their influence can be direct, such as those who fund the project, or indirect, such as end users who engage with the final product. For a project as encompassing as RWD, the stakeholder landscape is diverse. Let's look at the stakeholders involved in an RWD project in the following list:

- **Executive leadership**: At the top of the chain, executive leaders and decision-makers need to understand the business value of RWD. Their buy-in is crucial, not just in terms of budgetary allocations but also in steering organizational culture toward digital adaptability.

- **Marketing and branding teams**: The digital face of a brand has a significant bearing on its market perception. The marketing and branding teams, thus, have a vested interest in ensuring the brand's digital presence is consistent, impactful, and resonates across devices.

- **Sales and business development**: In an era where B2B and B2C interactions frequently begin online, the sales team's performance can be directly impacted by the usability and responsiveness of digital touchpoints.

- **IT and development teams**: Beyond the immediate team implementing RWD, the broader IT department has a stake in ensuring integrations, backend support, and infrastructure alignment with the new design principles.

- **End users**: While not typically considered in traditional stakeholder mapping, end users are the ultimate recipients of the RWD strategy. Their feedback, preferences, and pain points should be central to shaping the approach.

- **Supply chain partners**: For businesses in e-commerce or those with intricate digital supply chain integrations, partners and third-party vendors become crucial stakeholders. Their systems must mesh seamlessly with the responsive environment.

Mapping these stakeholders is only the first step. Engaging them, understanding their unique challenges and perspectives, and integrating their feedback form the bedrock of a holistic RWD strategy. The importance of this engagement cannot be understated. Misalignments or misunderstandings can lead to design revisions, escalating costs, and missed opportunities.

Moreover, by involving stakeholders early and often, organizations can foster a sense of collective ownership. This collaborative approach not only smoothens the RWD implementation process but also ensures that when the new design goes live, it's met with organizational enthusiasm rather than resistance.

In essence, while the technological aspects of RWD are undeniably critical, the human element, represented by these varied stakeholders, forms the heart of its success. Balancing their interests, managing their expectations, and harnessing their insights are integral to aligning RWD with the broader business landscape.

Next, we'll explore how to set goals that can realistically be achieved by businesses.

Setting measurable business goals

Success in business isn't just about lofty ideals or broad objectives. It's about setting clear, actionable, and measurable goals. As the adage goes, *What gets measured gets managed*. This principle becomes even more pivotal when we delve into the realm of RWD. Given the multifaceted nature of RWD, straddling technology, design, UX, and business outcomes, how does one set specific goals that can be tracked and evaluated? Here are some ways to do it:

- **The framework of SMART goals**: A time-tested method across industries, **SMART** goals – **Specific, Measurable, Achievable, Relevant, and Time-bound** – offer a robust starting point. By ensuring each RWD-related goal checks these boxes, companies can align their design aspirations with tangible business objectives.

- **Specificity is key**: Instead of setting a broad target, such as *Improve website UX*, setting a specific goal, such as *Reduce mobile page load time to under 3 seconds*, might be beneficial. This gives design and development teams a clear benchmark to work toward.

- **Quantify success**: Measurable goals allow for objective assessment; for instance, *Increase mobile traffic by 20% over the next quarter* or *Achieve a 15% uplift in conversions from tablet users*.

- **Reality check**: While ambition is commendable, setting unrealistic goals can be detrimental. It's vital to ensure the goals are achievable and rooted in research, industry benchmarks, and organizational capabilities.

- **Stay relevant**: Every goal should tie back to overarching business objectives. If increased user engagement on mobile devices is the aim, then RWD goals should center on mobile UX enhancements, content adaptability, and touch-friendly navigation.

- **Timelines matter**: Without a deadline, even the most well-intentioned goal remains nebulous. Time-bound goals, such as *Implement RWD changes before the holiday shopping season*, instill urgency, and foster accountability.

But setting SMART goals is just the beginning. To embed these objectives within the RWD strategy, you need the following:

- **Data-driven insights**: Harness analytics tools to gauge current website performance across devices. Understand user behavior, drop-off points, and engagement metrics. This data will not only inform goal setting but also provide a baseline for future comparisons.

- **Stakeholder alignment**: As discussed earlier, ensuring all key stakeholders understand and commit to these goals is crucial. Their feedback can also refine the objectives, making them more comprehensive and aligned with ground realities.

- **Iterative reviews**: The digital landscape is ever-evolving. Regularly revisiting set goals, measuring progress, and recalibrating based on results and new industry trends keeps the RWD strategy agile and relevant.

- **Feedback loops**: Engage real users through surveys, focus groups, or beta testing to gather feedback on RWD changes. Their firsthand insights can validate if the set goals genuinely enhance their browsing experience.

In conclusion, while RWD inherently focuses on adaptability and fluidity, its strategic implementation thrives on structure and precision. By setting measurable business goals, companies can navigate the complexities of RWD with clarity, ensuring that every design decision propels them closer to their desired business outcomes.

Now, how can we make use of UX to create more business?

Using UX as a business driver

The digital age has heralded a new wave of consumer empowerment. In this dynamic landscape, the relationship between businesses and their users isn't just transactional; it's experiential. The linchpin? UX. UX, especially in the context of RWD, isn't just about aesthetics or functionality. It's an intricate tapestry woven with business objectives, brand values, and user aspirations. Let's explore how UX becomes a powerful driver in achieving business goals and why it's so intrinsically linked with RWD:

- **The paradigm shift**: Gone are the days when a visually appealing website sufficed. Today's discerning users demand websites that not only look good but feel good. They expect intuitive navigation, seamless interactions, and quick information retrieval, irrespective of the device they're using. This shift isn't just about catering to user whims; it's a business imperative. A positive UX can lead to enhanced brand loyalty, higher conversions, and competitive differentiation.

- **UX and business KPIs**: An effectively designed UX can dramatically influence KPIs. Consider the bounce rate. If users find a website cumbersome or non-intuitive on their device, they're more likely to leave without taking desired actions, impacting metrics such as conversion rates and average session durations. Conversely, a seamless UX can boost these KPIs, directly enhancing revenue streams.

- **Personalization and engagement**: One of the cornerstones of modern UX is personalization. Users no longer want generic content. They seek curated experiences tailored to their preferences and behaviors. With RWD, businesses can create adaptive content strategies, ensuring users get relevant content, which in turn boosts engagement and fosters loyalty.

- **Trust and credibility**: A responsive website that offers a stellar UX isn't just a technical marvel; it's a brand statement. It communicates that the business values its users, understands their needs, and is technologically adept. This boosts brand credibility and instills trust, a crucial factor, especially in sectors such as e-commerce or fintech.

- **Feedback and iteration**: An optimal UX isn't a one-time achievement; it's an ongoing endeavor. Businesses should actively seek user feedback, conduct usability testing, and iterate the design based on insights. This feedback loop ensures that the UX remains relevant, aligning with evolving user preferences and emerging tech trends.

- **Holistic branding**: Every touchpoint, be it a product page or a checkout process, communicates brand values. By ensuring consistency and responsiveness across devices, businesses can convey a unified brand narrative, enhancing brand recall and equity.

- **The ROI of good UX**: Investing in UX isn't an expense; it's an investment. The returns, while sometimes intangible in the short term, can be significant. A superior UX can lead to reduced customer acquisition costs, higher **lifetime value** (**LV**), and amplified word-of-mouth marketing.

To encapsulate, in a world where digital interfaces are often the first (and sometimes the only) touchpoint between businesses and their users, UX takes center stage. Through RWD, businesses can ensure that this experience remains consistent, delightful, and aligned with business objectives, irrespective of the device. Remember — in the digital realm, your UX is your brand, and it's time businesses recognized and leveraged its profound potential.

Economic impacts of RWD

In today's digital era, the significance of a website extends beyond being a mere online presence. It's a powerful business tool with potential economic implications, both positive and negative. As business landscapes evolve and consumer behavior shifts toward mobile-first, RWD emerges not just as a design choice but as a strategic investment. This investment holds the promise of significant economic benefits for businesses willing to adapt. Let's unpack the economic impacts of RWD and understand why it's more than just a design approach:

- **Enhanced user engagement**: User engagement directly translates to potential revenue. With RWD, businesses ensure that users, regardless of their device, receive an optimal viewing experience. This decreases bounce rates and increases time spent on the site, leading to higher chances of conversion.

- **Boosted conversions**: A responsive design ensures that CTAs, product images, and checkout processes adapt smoothly to any screen size. This optimization removes barriers to conversion, resulting in increased sales, sign-ups, or any defined business action.

- **Positive SEO impacts**: Search engines, especially Google, prioritize mobile-friendly sites. An RWD site, being inherently mobile-optimized, stands a better chance of ranking higher in search results. Higher visibility in search equates to increased organic traffic, which, when coupled with an optimized UX, can lead to increased revenue.

- **Reduced development costs**: Developing separate sites for desktop and mobile can be resource-intensive. By opting for an RWD approach, businesses can consolidate their efforts into a single site that caters to all devices, leading to savings in development time and costs.

- **Broadened market reach**: As mobile internet usage continues to surge, RWD allows businesses to tap into a broader audience base. An adaptable website means being accessible to users at any time, anywhere, on any device — expanding market reach and potential revenue streams.

- **Decreased bounce rates**: Nothing deters potential customers faster than a website that's hard to navigate on their device. An unresponsive site can lead to higher bounce rates, meaning lost potential sales. RWD ensures users stay engaged, reducing bounce rates and the associated potential revenue loss.

- **Future scalability**: The digital landscape is ever-evolving. New devices with varying screen sizes will inevitably emerge. Investing in RWD ensures a level of futureproofing, allowing businesses to adapt to new devices without significant overhauls, thus ensuring sustained economic benefits.

In conclusion, the economic implications of RWD are profound. It's not just about aesthetics or UX — though those are undeniably important. At its core, RWD is an economic driver — a tool that, when wielded strategically, can lead to tangible business growth in the digital age. For businesses striving to maximize their digital ROI, understanding and harnessing the economic potential of RWD is crucial.

Beyond that, making sure that your strategy for RWD is consolidated is also a key aspect.

RWD in an omnichannel strategy

In a modern digital landscape marked by an abundance of devices, platforms, and touchpoints, businesses face the daunting challenge of offering a consistent and seamless UX. This is where an omnichannel approach comes into play, aiming to provide customers with a unified brand experience, whether they're interacting with a business through a mobile app, desktop website, physical store, or any other channel. Integrating RWD within this strategy becomes paramount. Let's explore how RWD amplifies the efficacy of an omnichannel approach and why it's an essential component for businesses:

- **Unified brand experience**: At the heart of an effective omnichannel strategy lies consistency. Customers should feel the brand's essence, irrespective of the channel they choose. RWD ensures that websites and online platforms automatically adjust to any device's screen size, ensuring that the brand's visual identity, message, and overall experience remain consistent.

- **Seamless transition between devices**: As consumers often switch devices during their purchase journey — starting on mobile, continuing on a desktop, and perhaps concluding on a tablet — RWD ensures that they receive a seamless experience throughout. This smooth transition fosters trust and reduces potential friction points in the customer's journey.

- **Centralized content management**: With an RWD approach, businesses can manage content centrally for various devices. This not only reduces operational inefficiencies but also ensures that all channels are updated simultaneously, guaranteeing that customers always receive the most current and relevant information.

- **Improved data collection and analysis**: An integrated RWD and omnichannel strategy allows businesses to collate data from various touchpoints into a unified system. This holistic view of customer interactions offers deeper insights into behavior patterns, preferences, and pain points, enabling businesses to refine their strategies more effectively.

- **Increased engagement opportunities**: In an omnichannel world, every touchpoint is an opportunity to engage. By ensuring that websites and platforms are responsive, businesses maximize these opportunities, ensuring that customers remain engaged, whether they're browsing on their phones during a commute or exploring products on a desktop at home.

- **Enhancing offline with online**: RWD plays a pivotal role in blending the online and offline worlds. A customer might browse products on a responsive site and then choose to visit a physical store for purchase. The cohesive experience between the responsive site and the store solidifies the brand's reliability in the customer's mind.

- **Future readiness**: As the digital sphere continues to evolve with new devices and touchpoints, RWD ensures that businesses remain ready for future challenges. It provides the flexibility needed to adapt to emerging channels, ensuring that the omnichannel strategy remains robust and effective.

In wrapping up, integrating RWD into an omnichannel strategy isn't just a best practice — it's a business imperative. As the lines between online and offline continue to blur, RWD acts as the glue binding various touchpoints, ensuring that businesses deliver a consistent, engaging, and reliable experience to their customers across the board. In the quest for true omnichannel excellence, RWD stands as a vital ally.

When done well, the omnichannel approach will help give you an edge in the market.

Competitive advantage and market differentiation

In today's competitive digital landscape, businesses continuously seek avenues to distinguish themselves, aiming not only to capture but also to retain consumer attention. Amid this, RWD emerges not just as a tool for aesthetic appeal or functionality but as a significant contributor to competitive advantage and market differentiation. Here's how adopting a robust RWD strategy can position a business distinctively, setting it ahead of the curve:

- **Elevating UX**: First impressions can often be lasting. When users access a site that adjusts seamlessly to their device, offering fluid navigation and impeccable visuals, they're more inclined to view the brand favorably. This optimized UX becomes an immediate differentiator, especially when competitors lag in offering a similar level of digital finesse.

- **Boosting organic discoverability**: With search engines such as Google prioritizing mobile-friendly sites, a responsive design can significantly influence a brand's search engine ranking. In a world where consumers often start their purchase journey with a search query, higher visibility can translate to increased traffic, engagements, and conversions, giving businesses an edge.

- **Catering to diverse audiences**: As the device ecosystem continues to evolve, from smartphones and tablets to wearables and beyond, RWD ensures a business is accessible to its audience, irrespective of their device preference. This universality in access can widen a brand's reach, attracting diverse segments of consumers.

- **Reducing operational redundancies**: Managing multiple versions of a site — for desktop, mobile, and tablet — can be operationally taxing and riddled with inefficiencies. RWD simplifies this, allowing for a centralized content and design management system. The saved resources can then be invested in other strategic initiatives, amplifying the business's competitive positioning.

- **Enhancing brand perception**: In the digital age, a business's website often serves as a primary touchpoint, influencing consumer perceptions. A responsive site not only reflects technical prowess but also demonstrates a brand's commitment to offering superior UXs. Such implicit messaging can significantly elevate brand reputation, making it a preferred choice amid competition.

- **Futureproofing the business**: The digital realm is in constant flux, with innovations cropping up regularly. An RWD approach ensures that businesses are primed for future adaptability, whether it's catering to new device formats or integrating emerging technologies. This proactive stance can be a compelling differentiator, signaling to consumers that the brand is ahead of its time.

- **Building trust and credibility**: Users, especially in e-commerce settings, seek reassurance about a site's credibility before initiating transactions. A well-implemented RWD not only reduces glitches and functional errors but also offers a polished, professional look, reinforcing trust and enhancing credibility in the eyes of potential customers.

Objectively, all of these advantages are imperative for a business to have. It's unique that a single concept, such as RWD, could provide that to a business.

But how do we ensure that we keep those advantages for the long term?

Ensuring long-term scalability

In an ever-evolving digital landscape, businesses often grapple with two seemingly opposing needs: delivering a solution fit for the present while also laying the groundwork for the future. The challenge intensifies in web design, where rapidly changing user behaviors, technological advancements, and market dynamics can render today's cutting-edge solutions obsolete tomorrow. In this context, RWD emerges as an ally, instrumental in ensuring that organizations can scale and evolve without perpetually revisiting their foundational digital strategies.

Let's delve into how RWD aids in futureproofing businesses. Here are the ways:

- **Flexibility amid change**: At its core, RWD is about adaptability. Whether it's accommodating new screen sizes, adjusting to novel user behaviors, or integrating fresh functionalities, responsive design principles enable a smooth transition. This inherent flexibility means businesses can adapt without overhauling their digital presence continually.

- **Reduced tech debt**: In the world of software and web development, technical debt refers to the future costs associated with bypassing the best approach now for a quicker, more expedient solution. By choosing a responsive framework from the outset, organizations mitigate the potential accumulation of technical debt, averting the costly and time-intensive processes of reworking or redesigning down the line.

- **Seamless integration of innovations**: As technologies such as AR, VR, and AI grow more prevalent, websites will need to incorporate these elements to stay relevant. RWD offers a versatile foundation, allowing for the smoother integration of such innovative features and ensuring that businesses remain at the forefront of digital experiences.

- **Sustained user engagement**: Users' expectations evolve with the digital ecosystem. By ensuring a site remains consistently user-friendly across devices and platforms, businesses can retain and grow their user base, fostering sustained engagement and loyalty.

- **Resource efficiency**: From a business perspective, continuously redesigning a website to meet emerging standards isn't just operationally cumbersome but also resource-intensive. RWD, by championing a one-size-fits-all approach, ensures that design, testing, and deployment resources are efficiently utilized, freeing up bandwidth for other strategic initiatives.

- **Embracing the mobile future**: As mobile internet usage continues its upward trajectory, ensuring a website's scalability across mobile devices isn't just a best practice; it's a business imperative. RWD ensures that as mobile technologies evolve, businesses remain poised to harness their potential fully.

- **Staying ahead of the competition**: In a competitive marketplace, the ability to scale and evolve rapidly often differentiates industry leaders from followers. By ensuring long-term scalability through RWD, businesses can stay agile, responding promptly to market shifts and maintaining a competitive edge.

Becoming efficient and staying in touch with technology is key to ensuring market dominance.

That doesn't mean that it's always easy. There will be challenges. Let's take a look.

Challenges in implementing RWD

The foray into RWD introduces businesses to a vast and evolving digital expanse. RWD promises a world of opportunities but, as with any journey through unknown terrain, it brings its share of complexities.

The digital age has blessed us with a near-endless number of devices. From expansive desktop screens to diminutive displays of smartwatches, the range is enormous. However, RWD isn't just about squeezing content into varying screen sizes. It's a quest to consistently deliver an optimized UX across a multitude of platforms. Each device, with its unique specifications and user interaction paradigms, presents its own set of challenges.

Yet, the intricacies don't stop at devices. Consider the array of web browsers available — Chrome, Safari, Firefox, Edge, and others. Each one uses a distinct rendering engine. This results in subtle (and sometimes pronounced) variations in how web content appears and functions. If unaddressed, these inconsistencies can lead to fragmented UXs, undermining the very essence of RWD.

Add to this the dynamic nature of the digital realm. It's an environment where change is the only constant. As businesses grapple with current technologies, new devices emerge, user behaviors evolve, and fresh technologies ascend. Thus, a responsive strategy cannot afford to be stagnant. Agility is not just desired; it's imperative.

However, RWD is more than a mere design challenge; it intricately intertwines with content. How does a business ensure its message remains potent across devices? How can it decide which content elements take precedence when screen space is so coveted? It's this delicate balance between design and content that often poses challenges.

Performance is another crucial facet of RWD, often overshadowed by design adaptability. A site might adapt beautifully across devices, but if it's sluggish or drains mobile data, users will be quick to abandon it. Thus, achieving harmony between visual appeal and performance is paramount.

On the organizational front, RWD isn't just an IT project. It necessitates a harmonious collaboration across departments — from the creative minds in design to the strategists in marketing. Ensuring this alignment, particularly in larger entities, can be a formidable challenge in itself.

Finally, there's the challenge of expertise. While the market brims with web designers and developers, professionals who truly understand the nuances of RWD are a rarer find. This potential skills gap can impede the effective rollout of responsive strategies.

In essence, while RWD offers a realm of potential, the path is dotted with challenges. But as with any journey, understanding the landscape paves the way for a smoother journey, allowing businesses to harness RWD's transformative power effectively.

Screens are our biggest asset in RWD but also our biggest challenge.

Device and screen fragmentation

In the golden age of the internet, web designers largely catered to a handful of standard screen resolutions. However, as we ventured further into the 21st century, the digital landscape became increasingly fragmented. Today, the sheer variety of devices — each with its unique screen dimensions and resolutions — has grown exponentially. From desktop monitors to laptops, tablets to smartphones, smartwatches to smart TVs, the device ecosystem has never been more diverse.

This burgeoning array of devices poses a profound challenge: How does one design a website that looks appealing and functions optimally across all these varying screen sizes? It's a monumental task, especially when considering the nuanced differences in user interactions across these devices. A hover effect, for instance, which works seamlessly on a desktop with a mouse, loses its relevance on a touchscreen device.

And it's not just about the horizontal and vertical screen resolutions. There's the pixel density to consider. Devices with higher pixel densities, often termed *Retina displays* or something similar by manufacturers, can render sharper images. However, they also require higher-resolution graphics to prevent them from appearing pixelated. This consideration plays a pivotal role in image selection and optimization, ensuring graphics are crisp without unnecessarily burdening data loads.

The situation is further complicated by device orientation. While a user might primarily interact with a tablet in landscape mode, the same device can easily be flipped to portrait mode, instantly altering the design canvas. This fluidity necessitates designs that not only adapt to varying screen sizes but also orientations.

Yet, the fragmentation isn't restricted to physical screens. The software environment, primarily the browsers, brings its own set of challenges. A design that renders beautifully on Chrome might have elements breaking on Firefox or Safari. Keeping abreast of browser updates and ensuring compatibility can sometimes feel like a never-ending game of whack-a-mole.

Also, consider the user scenarios. A person using a desktop is likely to be in a different context compared to someone accessing the site from a smartphone while on the move. The urgency, attention span, and the very intent can vary dramatically based on the device in use.

In many ways, device and screen fragmentation embodies the quintessential challenge of RWD. It demands a delicate dance between flexibility and consistency, ensuring users receive an optimal experience irrespective of their choice of device. While this fragmentation can seem daunting, it also presents opportunities: opportunities to engage users in their preferred environments, to innovate in design and functionality, and to truly embrace the digital age in all its varied glory.

Optimal experiences will mean nothing without the performance to go along with it.

Performance and speed concerns

As the digital age flourishes, our expectations for online interactions have evolved. Speed and performance are no longer luxuries; they are foundational. An efficient website that loads quickly is the baseline, the minimum standard users expect. In the realm of RWD, these performance considerations take on added layers of complexity and importance.

Imagine this: You've crafted a beautiful, responsive website that seamlessly adapts to various screen sizes. But as soon as a user taps on a link, the site takes a painstakingly long time to load. The immersive design is rendered ineffective by sluggish performance. The outcome? A potential customer might just bounce off, possibly to a competitor's swifter website.

The challenge is multi-fold. Each device, from a high-end desktop to an entry-level smartphone, has different processing capabilities. What loads instantaneously on a powerful computer might stutter on a mobile phone. Moreover, diverse network conditions further amplify the issue. While someone in an urban center with high-speed broadband might enjoy a frictionless experience, another user in a remote area relying on 3G may not be as fortunate.

High-resolution images and dynamic elements, while visually captivating, can be bandwidth-heavy, leading to longer load times. Furthermore, RWD, by its very nature, requires additional code and frameworks to ensure adaptability across devices. If not optimized, this extra code can bog down performance.

Yet, it's not just about the actual load time. It's about perceived performance too. If a website feels slow, if interactions aren't snappy, users might perceive it as laggy, even if the actual load times are within acceptable bounds.

Mobile users, often on the go, are particularly sensitive to website performance. Research (`https://www.bidnamic.com/resources/how-website-speed-affects-your-conversion-rates`) suggests that sites taking longer than 3 seconds to load see a significant drop in conversion rates. In the hyper-competitive digital landscape, these few seconds can mean the difference between a sale and a lost opportunity.

Addressing performance concerns in RWD means looking at optimization holistically. It's about employing techniques such as lazy loading, where content is loaded only when necessary. It's about compressing images without compromising quality and utilizing **content delivery networks** (**CDNs**) to ensure faster delivery times. Code minification and browser caching can further streamline the experience.

But perhaps most importantly, it's about cultivating a mindset where performance is as integral to the design process as aesthetics. It requires constant monitoring, regular testing across devices, and an unwavering commitment to delivering the best possible UX. After all, in the online realm, speed isn't just a feature; it's a determinant of success.

Another key aspect of successful design is how your content is structured.

Content prioritization and structuring

The online realm is bustling with information. Every website, every page, seeks to convey a message, offer insights, or engage users in some manner. Amid this flood of data, how does one ensure that the core message isn't drowned? Enter the challenge of content prioritization and structuring in the context of RWD.

When crafting a website's design, it's tempting to showcase all that a brand has to offer, right at the forefront. However, on a smaller device or a constrained screen, this approach can quickly lead to clutter, leaving users overwhelmed and disoriented. RWD isn't just about shrinking content to fit screens but discerning what's pivotal and what's supplementary.

Think of it as curating an art exhibit. While every piece is invaluable, the curator selectively places each one, guiding visitors' attention and ensuring they grasp the essence without feeling overwhelmed. Similarly, in RWD, it's crucial to determine what content takes center stage and what can be navigated to with a scroll or a click.

Users typically follow an *F* pattern when browsing — starting at the top left, moving horizontally, then vertically down the page. Leveraging this knowledge can guide decisions about where to place high-priority content. It's not just about the *what*, but the *where*.

The mobile-first approach offers a unique perspective. By designing for the smallest screen first, one is naturally inclined to prioritize content. What is absolutely essential for the user to see? Once this is determined for mobile, scaling up for tablets or desktops becomes a process of adding rather than subtracting, ensuring the core message remains consistent.

This doesn't downplay the significance of supplementary content. It's about creating a hierarchy. Key messages, CTA buttons, and primary navigation elements take precedence. Meanwhile, secondary information can be nested under expandable menus or placed further down the page.

Moreover, structuring isn't just about placement. It's about presentation. How is the content delivered? Is it a banner, a slider, a video, or a text block? Effective RWD considers not just the content itself but its format. Videos, for instance, can be bandwidth-heavy and might not be ideal for mobile users on limited data plans. Alternatives such as infographics or animated GIFs can convey similar messages in a more data-friendly manner.

However, prioritization and structuring are dynamic processes. With the ever-evolving digital landscape and shifting user preferences, what's pivotal today might be secondary tomorrow. This necessitates continuous assessment, user feedback integration, and adaptability in content strategy.

In essence, content prioritization and structuring in RWD is an exercise in empathy. It's about understanding user needs, recognizing their browsing patterns, and crafting an experience that feels intuitive, informative, and engaging, irrespective of the device in hand.

Our next step is to determine the best path for users when using our platforms.

Navigational complexities

Navigating the digital corridors of today's websites can be likened to traversing a vast, sprawling metropolis. There are main roads, side alleys, and countless destinations. However, the experience can range from a leisurely stroll through well-marked lanes to a confusing maze with no clear direction. The challenge of navigational complexities in RWD is akin to urban planning in this digital cityscape.

At the heart of this challenge is a fundamental question: How can a site be designed in a way to ensure that users effortlessly find what they seek, irrespective of the device they're on? This query gains even more weight in the RWD paradigm, where the same content must be accessible across diverse screen sizes and device types.

A common misstep is to assume that what works for a desktop will naturally translate to a mobile interface. On a broad desktop screen, expansive menus with multiple dropdowns might seem intuitive. However, when compressed onto a mobile screen, the same menus can turn into a cluttered mess, impeding user navigation.

One solution that emerged is the "hamburger menu," those three horizontal lines we often see on mobile interfaces. While it consolidates menus efficiently, it isn't without its critics. Some argue it hides essential paths, making them less discoverable. Others believe it's now universally recognized and, therefore, effective. This ongoing debate underscores the nuances of navigational decisions in RWD.

Touch versus click is another layer to this challenge. While desktop navigation relies heavily on precise cursor movements, mobile navigation is about touch, swipe, and pinch actions. Buttons must be adequately sized for fingers, swipe gestures need to be smooth, and dropdowns must be touch-friendly. Ensuring that navigational elements cater to both interaction modes is pivotal for a consistent UX.

Scrolling behavior also varies. On desktops, users might be more inclined to scroll horizontally, given the widescreen layout. In contrast, vertical scrolling is more intuitive on mobile devices. Designing with these tendencies in mind is essential for RWD.

Another consideration is the depth of navigation. How many clicks or taps should it take a user to reach their destination? The general rule of thumb is to keep it minimal but without compromising the richness of content. Too shallow, and the site feels sparse; too deep, and users might abandon their journey out of frustration.

A practical solution many designers adopt is progressive disclosure — revealing information step by step, based on user actions. This approach ensures that users aren't overwhelmed while still having access to detailed content if they wish.

Lastly, user feedback is invaluable. Tools such as heatmaps, which show where users most frequently click, or user journey analytics can provide insights into navigational pain points. Incorporating such feedback ensures that RWD navigation evolves in tandem with user needs.

In summary, crafting navigational paths in RWD is a dance between design principles and user preferences. It's about ensuring that, whether a user is on a vast desktop boulevard or a narrow mobile alley, their journey is smooth, intuitive, and enjoyable.

Resource allocation and project management

In the realm of RWD, the blueprint isn't merely about pixels and aesthetics. It's a complex tapestry that weaves together various teams, each contributing to the bigger picture. From the initial design mock-ups to the lines of code making those designs a reality, and from content creation to quality assurance tests, the path to a responsive site is layered and intricate. At the crossroads of this journey lies the critical aspect of resource allocation and project management.

Consider the scale of a typical RWD project: Designers are busy crafting adaptable layouts. Developers are engaged in making those designs functional across various devices. Content creators are ensuring that the site's message is consistent and effective. And testers are probing for any anomalies or glitches. With so many wheels turning simultaneously, how does one ensure that resources — both human and technical — are used optimally?

The first step is setting clear objectives. What is the RWD project aiming to achieve? Is it a redesign of an existing site, or is it a fresh digital initiative? Clarifying the goals at the outset provides a roadmap for subsequent stages.

Once objectives are set, the next challenge is timeline management. RWD projects are notorious for their shifting goalposts, often due to unforeseen technical hurdles or design revisions. Establishing a realistic yet firm timeline, with milestones, ensures that all teams are in sync and working toward a common deadline.

But time isn't the only resource at play. Budget considerations are equally pivotal. Allocating funds judiciously across the project's needs — be it for specialized RWD tools, hiring external expertise, or investing in testing infrastructure — can be the difference between a project's success and its stagnation.

Effective communication is the glue that holds these elements together. Regular check-ins, feedback loops, and open channels of dialogue ensure that any bottlenecks are addressed promptly. Whether it's a designer needing clarification on a specific layout aspect or a developer flagging potential scalability concerns, open communication ensures that the project remains agile and adaptable.

Another layer to consider is prioritization. Not all features or elements in an RWD project carry the same weight. For instance, ensuring that the main **CTA** is prominently visible across devices might take precedence over making a tertiary sidebar menu responsive. Discerning what to prioritize, based on business objectives and user needs, can streamline the project's focus.

Tools and platforms, such as agile project management software, play an instrumental role in this orchestra. They offer a centralized view of the project's progress, allow for task assignments, and facilitate inter-team collaboration. Investing in such tools, tailored to RWD's specific needs, can significantly enhance productivity and oversight.

In conclusion, while the visible outcome of an RWD project is a seamlessly adaptive website, the invisible machinery powering this outcome is a meticulous blend of resource allocation and project management. It's a reminder that in the digital symphony of responsive design, every note, every chord, and every crescendo is the result of careful planning, coordination, and execution.

It's now time to discuss why it's crucial to always make sure you're staying updated and informed as the technological landscape continues to change.

Staying updated – evolving trends and technologies

Diving into the realm of RWD is akin to stepping into a river with a swift current. Just as water flows and changes course, technology and trends in RWD are constantly evolving, making it a dynamic field that keeps web designers and developers on their toes. For businesses looking to thrive in this digital age, understanding and adapting to these changes is not just beneficial — it's essential.

The inception of RWD was a response to a changing digital landscape. The rise of smartphones and tablets necessitated websites that could adapt and deliver optimal experiences irrespective of the device. However, this was just the beginning. As technology progressed, so did the demands and expectations of users. From wearable devices to smart TVs, the range of screens vying for a responsive design broadened.

One of the challenges is the rapid pace at which new devices enter the market. Each device, with its unique screen size and resolution, adds another layer of complexity to the RWD puzzle. For designers and developers, this means perpetually updating their knowledge and skills. Regularly attending workshops, webinars, and conferences becomes indispensable to stay in tune with the latest practices and tools.

But it's not just about hardware. The software ecosystem around RWD is bustling with innovation. New frameworks, libraries, and tools emerge, offering enhanced capabilities for creating responsive designs more efficiently and effectively. Staying updated with these tools can significantly influence the quality and speed of RWD projects.

Moreover, the expectations of web users have seen a shift. It's no longer enough for a site to simply fit different screens. Today's users demand smooth interactions, intuitive navigation, and rapid load times. Innovations such as **Progressive Web Apps** (**PWAs**) and **Accelerated Mobile Pages** (**AMP**) have risen to address these demands, emphasizing not just adaptability but also performance and UX.

On the horizon, we see the inklings of even more transformative trends. With the growth of AR and VR, how will RWD principles apply to these immersive experiences? The integration of AI and **machine learning** (**ML**) presents another frontier. Imagine responsive designs that not only adapt to devices but also to individual user behaviors and preferences in real-time.

However, with the promise of innovation also comes the risk of obsolescence. Techniques or tools that were deemed cutting-edge a year ago might become redundant today. This transient nature of RWD trends underlines the importance of continuous learning and adaptability for businesses and professionals in the field.

In essence, navigating the RWD landscape is a journey of continuous evolution. It demands vigilance, a proactive mindset, and the willingness to embrace change. By staying updated with evolving trends and technologies, businesses can ensure that their digital presence remains not just relevant but exemplary, catering to users today while being poised for the users of tomorrow.

And with that, we've come to the end of our first chapter.

Summary

As we conclude this chapter on RWD, it's essential to reflect on the key lessons and skills we've garnered. We started by exploring the evolution of RWD, understanding how it has transitioned from a technical design approach to a vital element in strategic business planning. You learned about the principles that underpin responsive design, including fluid grids, flexible images, and media queries, and how these elements work together to create seamless UXs across various devices.

We delved into the practicalities of implementing RWD, discussing strategies to evaluate and enhance your existing digital assets for responsiveness. You gained skills in analyzing current web designs, identifying areas for improvement, and implementing changes that ensure efficiency and accessibility. Furthermore, we covered the significance of RWD in improving search engine rankings and enhancing overall user engagement.

Why are these lessons and skills crucial? In an increasingly digital world, RWD is not just about aesthetics; it's about accessibility, user satisfaction, and, ultimately, business viability. The ability to create web content that adapts gracefully across a multitude of devices is no longer optional; it's imperative for reaching and engaging a diverse audience. The skills you've acquired here lay the groundwork for not just surviving but thriving in a digital landscape that values adaptability and user-centricity.

Looking ahead, our journey into the digital realm continues in the next chapter. Building on the foundations of RWD, we will shift our focus more specifically to the growing importance of mobile devices in web design and user interaction. You'll learn how adopting a mobile-first approach is not just a trend but a crucial aspect of modern web strategy, preparing you to meet the needs of an increasingly mobile-centric user base. This is the natural next step, taking the principles of RWD and applying them in a context where mobile devices are often the primary means of accessing the internet. Join me as we dive deeper into optimizing for the smallest screens to ensure the biggest impact.

2

Mobile-First Strategy in a Multi-Device World

In this chapter, we delve into the transformative world of a mobile-first strategy in a multi-device world. In today's digital age, the ubiquity of mobile devices has reshaped how we interact with the web, making a mobile-centric approach not just an option but a necessity for businesses and web developers. This chapter is designed to equip you with a deep understanding of the predominance of mobile in our digital landscape and the substantial ROI it can offer.

We will begin by exploring the current state of mobile usage, providing you with insights into why and how mobile has become the leading platform for digital interactions. You'll learn about the evolving user behavior patterns and preferences in a mobile-dominated world, emphasizing the need for a mobile-first approach in web design and development.

Understanding the ROI from a mobile-centric approach is crucial. We will guide you through the tangible benefits of prioritizing mobile, including increased engagement, improved user experience, and higher conversion rates. This section aims to demonstrate not only the immediate benefits but also the long-term value a mobile-first strategy can bring to your business.

Furthermore, we'll delve into how optimizing for mobile influences other platforms. A mobile-first approach doesn't mean neglecting desktop or other platforms; instead, it's about using the strengths of mobile optimization to enhance overall web strategy across all devices. You'll learn how a well-executed mobile strategy can lead to improvements in your web presence universally.

Finally, this chapter will address common pitfalls and challenges in mobile design. From navigating the intricacies of responsive design to ensuring optimal performance and usability on a variety of devices, we'll provide you with the knowledge to identify and overcome these hurdles effectively.

By the end of this chapter, you will not only understand the significance of a mobile-first approach in the current multi-device era but also possess the practical skills and insights needed to implement this strategy effectively in your projects. Let's embark on this journey to harness the power of mobile and transform your digital presence.

The topics covered in this chapter are as follows:

- Discussing the shift to mobile dominance
- Exploring the business advantages of mobile prioritization
- The ripple effect — from mobile to desktop
- Overcoming mobile design challenges

Discussing the shift to mobile dominance

We're part of a generation where technology isn't just a tool; it's an extension of our daily lives. Whether you're a tech enthusiast, a casual smartphone user, or somewhere in between, there's no denying the profound impact mobile technology has had on our lives. From a time when mobile phones were bulky devices reserved for a select few to the present day, where they serve as powerful mini-computers in our pockets, the journey has been nothing short of revolutionary.

Back in the day, mobile phones served a singular purpose — to make calls. Fast forward a bit and they incorporated the ability to send and receive text messages. It wasn't until the late 2000s, though, that mobile phones began their transformation into what we now recognize as smartphones. This transformation has been driven by two major factors: hardware evolution and software innovation.

On the hardware front, we've seen drastic improvements. Remember when a phone with a camera was a big deal? Or when "color screens" were the selling point? Nowadays, we have phones with multi-core processors, extensive **random-access memory** (**RAM**) capacities, and camera setups that rival the latest Canon or Nikon cameras. Such leaps in hardware capabilities have facilitated the rise of applications that demand more resources, from immersive games to advanced productivity tools.

On the software side, the emergence of platforms such as Android and iOS has given developers a playground, one where they are free to create, play, and innovate. These operating systems, coupled with some developer-friendly tools, birthed an era of mobile applications that truly changed the industry. No longer were phones just for calls and texts. They became our news sources, our gaming devices, our cameras, our maps, and even our personal assistants. Everything that we needed to get through our daily routines existed within the palms of our hands.

Then came the app stores, which provided developers with a platform to reach a global audience. This led to a rush of innovative apps that catered to virtually every need, from social networking to health tracking, and from online shopping to remote work solutions. It was this rush that solidified the relationship that phones now have with our personal and professional lives.

Yet, it's not just about smartphones. Over the past decade, we've seen the rise of tablets, smartwatches, and other wearable devices. These gadgets, while diverse in form and function, share a common thread — they're all designed to keep us connected and enhance our daily experiences.

But why does this matter? Simply put, the rapid evolution of mobile technology has fundamentally shifted the way users access and interact with digital content. It's this shift that businesses, developers, and designers need to recognize and adapt to. Because in today's digital era, understanding and leveraging mobile technology isn't just an option — it's a necessity.

By appreciating this evolution, we can better understand the *why* behind the dominance of mobile. It offers a lens through which to view the current landscape and provides insight into the trajectory of future trends. As we navigate through the intricacies of the mobile-first approach, it's essential to keep the journey of mobile technology, from its humble beginnings to its current prominence, in perspective.

Rise of the smartphone generation

Growing up, I recall a world of flip phones, keyboards that slid out, and the novelty of that first camera phone. But today, not even two decades later, those devices seem almost archaic, eclipsed by the exponential rise of the smartphone.

Think about it for a moment. Those of today's younger generation have grown up in a world where having a world of knowledge, instant communication, and endless entertainment at their fingertips is the norm. They don't know a world without app stores, touch screens, or instant messaging. They're the true digital natives.

It's not only about having access to information; it's also about the immediacy of it all. Why wait until you get home to search for something when you can just pull out your smartphone and get the answer instantly? Need to capture a moment? Gone are the days of disposable cameras or even digital cameras for most. We have high-resolution cameras right in our pockets.

But it's more than just convenience; it's about a cultural shift. Today, it's not uncommon to see groups of friends together, each engrossed in their own device. Or families at dinner tables, with younger members sneakily checking their phones under the table. For some, this is a sign of societal decay; for others, it's a new way of socializing. I'll leave you to decide which side of the fence you're on. Either way, it's undeniable evidence of the profound impact smartphones have had.

This generational shift, driven by the smartphone revolution, underscores the need for businesses, designers, and developers to prioritize mobile experiences. It's not just about being where the users are, but also about understanding how they think, interact, and engage in our now mobile-centric world.

And while there's a sense of nostalgia for the simpler days before smartphones took over, there's also an air of excitement for modern-day advancements. With technology continuously evolving, the smartphone generation is poised to lead the way, influencing how we all interact with the digital world for years to come. One thing's for sure: it has set the stage, and the world is eagerly watching to see what it'll do next.

The influence of the smartphone generation reaches deep into industries, affecting marketing strategies, retail habits, and even the design principles of digital interfaces. It has caused businesses to pivot their strategies, focusing more on mobile experiences. Not just because it's a growing medium, but also because it's where their audience is.

Changing user behaviors and expectations

In the dynamic digital landscape we navigate today, it's not just the technology that's shifting at breakneck speeds, but also how we, the users, interact with it. Our behaviors, molded by the constant exposure to evolving tech, are a testament to the profound influence of mobile devices.

Consider, for instance, the first time you used a smartphone. The sense of wonder at pinch-to-zoom or the convenience of an app that catered to your precise need. Fast forward to today, and these features aren't delightful surprises — they're baseline expectations. Our bar for what constitutes a smooth and intuitive mobile experience has been set incredibly high, and it's all thanks to the rapid advances in mobile technology.

As our daily routines became more intertwined with mobile usage, so did our impatience. We have been spoilt and have become a generation of instant gratifiers. Waiting more than a few seconds for an app to load or a page to refresh feels almost torturous. We expect notifications to be instant, updates to be timely, and any piece of information we desire to be available at our fingertips at a moment's notice.

But it's not just about speed. We crave personalized experiences. Algorithms have familiarized us with curated content. From personalized playlists on music apps to product recommendations when online shopping, users now anticipate a tailored digital experience that aligns with their preferences, history, and habits.

Moreover, our threshold for faulty or complicated designs is nearly nonexistent. If an app is not intuitive or if a website doesn't resize well on a mobile screen, chances are we'll abandon it pretty quickly. In a world abundant with options, loyalty is hard-won and easily lost.

So, what does this mean for businesses and developers? Firstly, there's no *one-size-fits-all* solution anymore. Users expect platforms to be optimized for their device, whether it's a smartphone, tablet, smartwatch, or any other connected gadget. Secondly, a seamless user experience isn't a luxury; it's the bare minimum. Any hiccups in navigation, long load times, or poor interactions can lead to a lost user or sale.

In essence, as mobile continues its dominance in our lives, the expectation is on businesses and developers to not just keep pace with changing user behaviors but to anticipate them. Only by staying a step ahead can they truly cater to the evolved and ever-evolving expectations of the mobile-savvy user.

Mobile's impact on online traffic

We've all been there: waiting in line at a café, commuting to work (don't text and drive), or even lounging on our couch during a lazy Sunday afternoon — all the while with our trusty smartphone in hand, effortlessly connecting us to the world. It's this pervasive nature of mobile devices that has fundamentally reshaped the way we access information, entertainment, and services online.

As we dive into the stats, it's mind-boggling to grasp the sheer volume of online traffic attributed to mobile devices. Recent data shows that over half of the global web traffic now originates from mobile. In some regions, especially where mobile connectivity has leapfrogged traditional desktop access, this figure skews even higher.

So, why is this shift significant?

For starters, this transition signals a change in user preference. The convenience of on-the-go browsing, coupled with the rapid advancement of mobile networks such as 5G, has solidified smartphones as the primary gateway to the digital realm. Their compactness, immediacy, and personal nature make them an obvious choice for quick searches, social media check-ins, or even in-depth research.

But it's not just about sheer numbers. The quality of traffic and user engagement patterns on mobile are distinct. Studies suggest mobile users often showcase a higher level of engagement thanks to the personal nature of smartphones. However, their browsing sessions might be shorter, reflecting the "snackable" consumption pattern where users frequently dip in and out of content.

The rise in mobile traffic has also led to the growth of mobile-first platforms and applications. Think of platforms such as Instagram or apps such as Snapchat — designed with mobile users at their core, leveraging the unique capabilities of smartphones, from cameras to location services.

Businesses, too, can't afford to remain bystanders in this shift. A surge in mobile traffic demands an optimized mobile user experience. Websites that are not mobile-friendly risk higher bounce rates, as users are likely to leave if they can't easily navigate or if the site loads too slowly. E-commerce platforms, in particular, have felt the ripple effects. With more consumers shopping via mobile, ensuring a seamless shopping and checkout experience has become paramount.

Yet, it's not just about optimization. This shift offers businesses a treasure trove of insights. Mobile traffic, with its unique user behaviors and patterns, can provide businesses with invaluable data. Understanding how users interact with content, what they're searching for, or even when they're most active can drive more targeted marketing strategies and product decisions.

In light of the mobile revolution's influence on online traffic, its mark on the digital landscape is undeniable. This shift not only prescribes new rules of engagement but also fosters the creation of innovative mobile-centric experiences. As mobile devices continue to be an integral part of our lives, it's evident that for businesses and developers alike, adapting to this transformation isn't merely advantageous — it's imperative.

Shifting business paradigms

In the wake of the mobile revolution, businesses have been presented with a clear and undeniable directive: adapt or become obsolete. The advent of mobile devices, especially smartphones, has altered not only how consumers engage with digital platforms but also how businesses conceptualize and execute their digital strategies.

Initially, the digital strategies of many businesses were primarily desktop-centric. Websites were built with the larger screens of laptops and desktops in mind, with mobile versions often being secondary or even an afterthought. However, as the number of mobile users surged, this desktop-only approach quickly proved to be outdated and ineffective. Mobile users sought fluid, intuitive experiences tailored to their devices, and businesses that failed to provide this often found themselves falling behind their more adaptive competitors.

Furthermore, mobile devices have introduced a myriad of new touchpoints for businesses to engage with their customers. Applications, push notifications, location-based services, and augmented reality experiences are just a few examples of the new tools that businesses have at their disposal. But with these opportunities come challenges, including the need for seamless integration between platforms and ensuring data privacy and security.

One clear indication of the paradigm shift can be observed in the realm of e-commerce. Previously, online shopping was primarily a desktop activity. But with the convenience of smartphones, a significant portion of consumers now prefer to shop on mobile apps. This change in consumer behavior prompted businesses to innovate, leading to the rise of mobile wallets, one-click purchases, and interactive product visualizations optimized for mobile screens.

Additionally, the emergence of mobile has democratized business opportunities. Smaller businesses that may have struggled with establishing a physical presence can now compete in the digital realm through effective mobile strategies. They can reach out to a global audience, offer personalized deals, and build loyalty programs, all tailored for the mobile user.

In essence, the rise of mobile technology hasn't just been about shifting from desktop to mobile screens; it's been a profound transformation in how businesses conceptualize and interact with their audiences. In this new paradigm, understanding the mobile user and their preferences, behaviors, and pain points is pivotal. Those businesses that can effectively navigate this new landscape will find themselves at the forefront of the next wave of digital innovation.

Exploring the business advantages of mobile prioritization

In today's fast-paced digital landscape, the adage "time is money" has never been more pertinent. For businesses looking to thrive in this environment, the capability to reach and engage with consumers instantly is no longer a luxury — it's an imperative. And that's where mobile prioritization shines brightest.

Imagine the life of an average individual today. Amid the hustle and bustle, their smartphone becomes more than just a device — it's an extension of them. Morning alarms, news updates, social media scrolling, online shopping, evening reads — everything happens there. This omnipresence of mobile devices in daily routines offers businesses an unprecedented opportunity: immediate accessibility to their audience.

Prioritizing mobile means that a business is always just a pocket-reach away from its potential customers. It allows for instant communication, whether that's through an app notification about a flash sale, a personalized email about a restocked item, or a location-based offer when they're in proximity to a store. This immediacy translates to increased touchpoints, offering businesses more opportunities to convert interactions into transactions. More touchpoints, more transactions.

Beyond just the speed of access, mobile prioritization enhances the breadth of reach. With over half of global web traffic now coming from mobile devices, a mobile-optimized platform amplifies a business's potential audience manifold. It breaks geographical barriers, allowing a local store in Cape Town to engage a customer in New York, all thanks to a mobile-friendly online storefront.

This immediate reach also has profound implications for data collection and user understanding. Each interaction on a mobile device — be it a product search, an app download, or a webpage visit — offers businesses valuable insights. These analytics can be harnessed to refine marketing strategies, tailor product offerings, and enhance user experience, ensuring that businesses are not just reaching their audience but resonating with them.

In a world where immediacy is key, businesses that can effectively tap into the power of mobile stand to gain not just in terms of sales, but also in building lasting, real-time relationships with their audience. However, while the advantages of immediate accessibility and reach are manifold, they also set the bar high for businesses. The modern mobile user, accustomed to the speed and convenience of their device, has little patience for subpar mobile experiences. Slow-loading pages, non-intuitive interfaces, or irrelevant notifications can quickly deter potential customers. Thus, while the mobile realm offers immense potential, it demands an equal measure of commitment to delivering quality and value at every touchpoint.

Enhanced user engagement on mobile

The transition to mobile-first strategies is not solely about adjusting to changing traffic sources or even meeting users where they are. One of the most compelling reasons to prioritize mobile is the significant potential it offers in enhancing user engagement. Here's a closer look at why mobile isn't just about reach — it's also about deepening connections.

Consider the personal nature of a mobile device. Unlike desktops or even laptops, smartphones are personal gadgets that accompany users throughout their day, from waking moments to bedtime. This personal connection between users and their devices creates an intimacy that businesses can leverage. Mobile devices aren't just tools; they're experiences, companions, and gateways to the digital realm.

This connection offers businesses an unparalleled opportunity to engage users in richer, more meaningful ways. Push notifications, for instance, aren't merely reminders; they can be tailored based on user behavior, preferences, and location. Imagine getting a notification about a discount at your favorite café just as you pass by it. This isn't just engagement — it's almost serendipitous interaction, all thanks to technology.

App interfaces, another mobile-specific feature, are tailor-made for enhancing user engagement. Unlike websites, which can sometimes feel generic, apps can be personalized to offer unique user experiences. This includes user-specific recommendations, loyalty rewards, and even gamification elements that make interaction fun and rewarding.

Moreover, the tactile nature of mobile devices adds another layer of engagement. Swiping, pinching, pulling — all these gestures make the interaction more visceral, creating a sense of involvement that's hard to replicate on other devices.

However, with these potential advantages come challenges. Enhanced engagement means that businesses need to be constantly attuned to their users' needs and preferences. A poorly timed notification or an intrusive ad can quickly sour a user's experience. And in the realm of mobile, where alternatives are just a tap away, user patience is thinner than ever before. It's a tightrope walk between engagement and intrusion.

Yet, for businesses that get it right, the rewards are immense. Engaged users are more likely to become loyal customers, offer feedback, and even become brand advocates. They spend more time on apps, interact more frequently, and are more open to discovering new features or products.

Prioritizing mobile isn't just a strategic shift — it's also a commitment to deeper, more enriching user engagement. It's about recognizing the evolving digital landscape and the unique opportunities that mobile offers in connecting with users. For businesses looking to foster lasting relationships in today's digital age, mobile offers a path filled with promise.

Optimized conversion rates

Mobile devices have transformed the way consumers interact with businesses, reshaping the digital marketplace landscape. Beyond accessibility and engagement, another significant benefit of a mobile-first approach is its potential to optimize conversion rates. When businesses align their strategies with mobile user behaviors and preferences, they're positioning themselves for increased success in driving users to desired actions.

The relationship between mobile optimization and conversion rates is straightforward: as users find it easier to navigate, understand, and interact with content on their mobile devices, the likelihood of them completing a desired action, whether it's signing up for a newsletter or making a purchase, increases.

The following features make mobile prioritization a game-changer for conversions:

- **Streamlined user experience**: On mobile, there's limited screen real estate, pushing designers and developers to prioritize essential elements and eliminate unnecessary fluff. This often leads to a more focused and straightforward user journey, guiding users effortlessly toward conversions.

- **Mobile-specific features**: Features such as click-to-call buttons, location-based offers, and one-tap checkouts take advantage of mobile device capabilities and streamline the conversion process. Such mobile-centric tools can drastically shorten the path to conversion, eliminating barriers that might deter a potential customer.

- **Faster loading times**: A mobile-optimized site is generally leaner than its desktop counterpart. Faster loading times not only improve user experience but directly correlate with higher conversion rates. Even a one-second delay can result in a significant drop in conversions, highlighting the need for speed in the mobile realm.

- **Personalized content delivery**: Mobile devices, with their array of sensors and data, can provide businesses with insights into user behaviors and preferences. This allows for more personalized content delivery, presenting users with products, services, or information that are most relevant to them, thus increasing the likelihood of a conversion.

- **Seamless integration with other mobile features**: Mobile sites and apps can be integrated seamlessly with other apps or features on a device. For instance, integrating a mobile shopping site with mobile wallets or payment apps can make the checkout process more fluid, reducing cart abandonment rates.

It's essential to approach mobile optimization with a user-centric mindset. Overloading a mobile site with aggressive ads, popups, or unnecessary elements can have the opposite effect, frustrating users and pushing them away. Additionally, continually monitoring and analyzing mobile user behaviors can provide insights, allowing businesses to tweak and refine their strategies for even better conversion rates.

Personalization and user context

Users are no longer satisfied with one-size-fits-all experiences. They expect and appreciate personalized content that resonates with their preferences, needs, and context. Mobile devices, due to their inherent nature, provide a unique vantage point for businesses to offer such tailored experiences, amplifying user engagement and, subsequently, business outcomes.

Mobile prioritization enables heightened personalization and understanding of user context through the following techniques:

- **Real-time data collection**: Mobile devices, with their myriad of sensors and functionalities, offer a treasure trove of real-time data. Whether it's location data, usage patterns, or even ambient light and sound, these data points can be harnessed to deliver content that's immediately relevant to the user's current situation.

- **Integration with native apps**: Mobile devices house a plethora of apps that can offer insights into a user's habits, preferences, and behaviors. Integration with these apps, whether it's a calendar, fitness tracker, or shopping app, can allow businesses to understand and cater to user needs more effectively.

- **Location-based services**: Geolocation capabilities allow businesses to serve location-specific content or offers. Imagine walking by a coffee shop and receiving a discount notification for your favorite drink. Such timely and relevant interactions foster loyalty and increase conversion rates.

- **Adaptive user interfaces**: Mobile-first designs can adapt dynamically based on user behavior and preferences. For instance, if a user frequently uses a particular feature or section of an app, it can be made more accessible, creating a tailored user interface that evolves with the user's habits.

- **Behavioral personalization**: Over time, by analyzing a user's interactions, search queries, and in-app behaviors, mobile platforms can predict what the user might be interested in next, offering suggestions or content that aligns with their anticipated needs.

- **Cross-device synchronization**: Many users switch between devices throughout their day. Mobile platforms that sync user data and preferences across devices can offer a consistent, personalized experience, irrespective of the device being used, ensuring a seamless user journey.

- **Contextual notifications**: Push notifications, when used judiciously, can serve highly relevant information based on the user's context. Whether it's a reminder to complete a task when they're in a particular location or a news update relevant to their interests, these notifications can enhance user engagement significantly.

It's vital, however, to strike a balance. Over-personalization or misuse of user data can come across as intrusive, leading to distrust and potential loss of users. Transparency in data usage, giving users control over their data, and ensuring data security is paramount.

In a nutshell, the mobile-first approach, when executed thoughtfully, opens the door to unparalleled personalization opportunities. By understanding and respecting the user's context, businesses can forge deeper connections, enhancing user loyalty and driving superior business results.

Building brand loyalty in the mobile era

The mobile era has transformed not only the way businesses operate but also the way they interact with and retain their customers. Today, brand loyalty isn't just about offering top-notch products or services; it's also about delivering consistent, personalized, and seamless experiences across the mobile landscape. Let's explore how a mobile-first strategy plays an instrumental role in cementing brand loyalty:

- **Constant presence**: Mobile devices, by nature, are almost always within arm's reach of their users. This omnipresence offers businesses a 24/7 touchpoint, ensuring they're always connected with their customer base. This consistent engagement fosters familiarity and trust.

- **Availability**: The ability to address user needs immediately, be it through a quick response on a chat app, a timely push notification, or real-time problem resolution, helps businesses earn trust and loyalty.

- **Personalized experiences**: As mentioned earlier, mobile platforms enable hyper-personalization. Tailored content, product recommendations, and even user interface adjustments based on individual behaviors can make users feel valued, enhancing their connection with the brand.

- **Unified omnichannel experience**: With mobile prioritization, brands can offer a consistent experience across all touchpoints, whether it's an app, a website, or even an offline store. When users receive a harmonized experience, regardless of where they interact with the brand, it reinforces brand reliability.

- **Loyalty programs and rewards**: Mobile platforms allow brands to introduce digital loyalty programs with ease. These programs, when integrated seamlessly with mobile apps or websites, can incentivize repeated interactions and purchases, strengthening long-term loyalty.

- **Community building**: Mobile-first strategies can empower businesses to foster communities around their brand. Be it through in-app forums, integrated social media feeds, or user-generated content, creating spaces for users to interact, share, and belong can amplify brand allegiance.

- **Empathetic support**: Mobile platforms enable businesses to offer on-the-go support. Immediate access to help centers, chatbots, and even human customer service representatives directly from mobile apps can reassure users, making them feel supported and valued.

- **Innovative mobile-only features**: Introducing features exclusive to mobile platforms, such as augmented reality shopping experiences and location-based offers, can give users reasons to stay engaged and loyal.

- **Feedback loops**: The mobile-first approach allows businesses to continuously gather feedback. Whether it's through app ratings, direct surveys, or feedback forms, understanding and acting on user insights can demonstrate a brand's commitment to its users.

To truly foster brand loyalty in the mobile era, businesses must ensure that their strategies are user-centric. It's not merely about being present on mobile devices but about creating genuine, meaningful interactions that resonate with users, making them not only customers but also ardent advocates of the brand.

Next, it's time to consider the evolution from desktop to mobile.

The ripple effect – from mobile to desktop

The digital landscape has undergone dramatic shifts in recent years, and the transition from desktop to mobile stands as a testament to this evolution. As the prominence of mobile began to rise, so did its influence on other digital platforms, especially desktops. This phenomenon, where changes in mobile user behaviors and expectations begin to influence and reshape the desktop experience, can be likened to the ripple effect observed in water. Just as a single drop can send waves across a still pond, innovations, and standards set in the mobile realm can propagate and bring about changes in the broader digital ecosystem.

In the context of web design and user experience, the ripple effect signifies the cyclical influence between mobile and desktop platforms. Early on, desktop websites were the standard, dictating design, function, and user experience principles. However, with the ubiquity of smartphones and their distinctive user interfaces, businesses quickly recognized the need for dedicated mobile designs. As these mobile designs evolved, they began to set new standards — standards that users came to expect not only on their phones but also when they transitioned back to their desktops.

One clear manifestation of this ripple effect is in the simplicity and intuitiveness of design, as seen in the chapter earlier. Mobile screens, due to their limited real estate, necessitated a decluttered, straightforward design approach. As users became accustomed to this streamlined experience on mobile, they began to desire similar simplicity on desktop websites. Consequently, we observed a trend where desktop interfaces started adopting the card layouts, hamburger menus, and minimalist designs that were initially mobile-centric conventions.

Another key area influenced by the ripple effect is page load times. Mobile users, often on the move, demanded fast-loading pages. This insistence on speed bled over to desktop users, who began to expect similarly swift load times regardless of the device they were on.

Moreover, the fluidity and responsiveness initially tailored to mobile devices began to set the standard for desktops. Users started anticipating a seamless experience, expecting websites to adjust flawlessly to varying screen sizes and resolutions, whether they were on a smartphone, tablet, or large-screen monitor.

Consistency across platforms

In today's multi-device digital landscape, users effortlessly shift between TVs, mobiles, tablets, desktops, and even smartwatches throughout their day. One moment, they might be browsing on their smartphone during a commute, and later, they could be continuing their exploration on a desktop at home. Given this seamless transition that is required across devices, one fundamental expectation arises: consistency.

Consistency, in the realm of web design and user experience, ensures that regardless of the device or platform, a user's interaction remains predictable, familiar, and intuitive. It's the bridge that allows users to carry their experiences, preferences, and expectations from one device to another without missing a beat.

The following reasons make consistency a critical element of user satisfaction:

- **User familiarity**: If a user learns how to perform a task on a website via their mobile, they expect to use the same knowledge to accomplish that task on a desktop. Consistent interfaces reduce the learning curve, making users feel competent and in control.

- **Trust and branding**: Inconsistencies, whether in design, functionality, or content, can undermine the user's trust in a platform. On the other hand, a consistent look and feel across devices reinforces brand identity, instilling a sense of reliability.

- **Efficiency and productivity**: For users who frequently switch between devices, consistent interface elements (such as buttons, menus, and layouts) mean they don't waste time searching or adapting to a new environment. They can pick up right where they left off.

- **Enhanced user satisfaction**: Users don't typically think about consistency until it's missing. When present, it offers a seamless and harmonious experience, leading to increased user satisfaction.

Achieving this consistency, however, requires a strategic approach. It isn't merely about making a mobile site look like its desktop counterpart. It's also about ensuring functional consistency. Color schemes, fonts, interactive elements, and even the tone of content play a part in weaving a consistent narrative.

While the emphasis has been on maintaining uniformity, it's also essential to understand that each platform has unique strengths and characteristics. Striking a balance between leveraging these unique features and maintaining cross-platform consistency is the golden mean that developers and designers should aim for.

As the lines between mobile and desktop continue to blur, and as users move fluidly between them, ensuring consistency is no longer a luxury — it's a necessity. In doing so, businesses not only offer a cohesive user experience but also pave the way for enhanced user engagement and loyalty.

Transferring mobile innovations

As we delve deeper into the multi-device era, it's not just about translating consistent experiences across platforms; it's also about recognizing and transferring innovations that originated in the mobile sphere to broader ecosystems, such as the desktop environment.

Mobile devices, due to their inherent constraints and unique opportunities, have been the birthplace of numerous innovations. Whether through touch gestures, device orientation-based interactions, or location-based functionalities, mobile has continuously pushed the envelope of what's possible in the digital realm.

The following functionalities carry out the crucial process of effectively transferring these mobile-born innovations to desktop and other platforms:

- **Adapting to user expectations**: Users who experience and adapt to innovations on mobile naturally come to expect these functionalities elsewhere. For instance, those accustomed to swiping on mobile may intuitively attempt similar gestures on touch-enabled desktop screens. Recognizing and addressing these expectations can lead to a more intuitive and delightful user experience.

- **Enhancing desktop functionalities**: While desktops have traditionally been seen as the "powerhouses" of computing, certain mobile innovations can significantly enhance their usability. Consider the adoption of app-based interfaces in desktop operating systems, inspired by mobile operating systems. This integration offers users a more streamlined and focused experience, mirroring their mobile interactions.

- **Driving uniformity in innovation**: While it's essential to maintain platform-specific advantages, there's merit in ensuring that groundbreaking functionalities find their way across the digital landscape. This strategy fosters an environment where innovation isn't siloed but is shared and propagated, leading to an overall richer digital ecosystem.

- **Fostering continuous improvement**: By continually assessing mobile innovations and integrating them into the desktop space, developers and designers are nudged into a cycle of perpetual improvement. This approach ensures that platforms don't stagnate but evolve in tandem, borrowing and lending features that enhance user experience.

However, this transfer isn't without challenges. Not every mobile innovation will seamlessly fit into the desktop mold. The key is discernment. It's crucial to evaluate whether a particular feature or functionality adds genuine value to the desktop environment or whether its inclusion would merely be a superficial gimmick.

In conclusion, as the ripple effects of mobile innovations permeate the broader digital world, they bring with them fresh perspectives, challenges, and opportunities. Embracing this flow, rather than resisting it, positions businesses at the forefront of user-centric design, ensuring they remain relevant, adaptive, and continually poised for the next wave of digital evolution.

Responsive design — a two-way street

In the context of our digitally connected world, it's tempting to think of "responsive design" merely as a way to make desktop websites fit neatly into the smaller screens of mobile devices. However, as we evolve further into the realm of multi-device experiences, it's becoming evident that responsive design is not just a one-way journey from big to small. It's a dynamic two-way street, bridging gaps and ensuring seamless experiences.

Traditionally, responsive design was often seen as a solution to a problem: "How do we make this desktop site look good on a smartphone?" However, as mobile devices became more prevalent and sometimes even the primary way users accessed the internet, the narrative began to shift. Suddenly, the question became: "How can our mobile-first approach also cater to desktop users effectively?" The following approaches can answer this question:

- **Mobile-up, not just desktop-down**: With a significant number of users experiencing the web for the first time through their mobile devices, businesses recognized the need to think mobile-up rather than just desktop-down. This shift meant that design considerations began with mobile constraints and then expanded outward to encompass larger screens, instead of merely shrinking a desktop design.

- **Leveraging mobile strengths**: As designers and developers approached responsive design from a mobile-first perspective, they found opportunities to harness the strengths of mobile devices. Features such as location awareness, touch gestures, and device orientation began to be creatively integrated, even in desktop experiences where possible.

- **Maintaining the core experience across devices**: The two-way street of responsive design ensures that the core user experience remains consistent regardless of the device. While specific functionalities might differ due to device capabilities, the essence of the interaction and the brand message remain consistent.

- **Adaptable grids and flexible elements**: A genuine responsive approach, be it mobile-to-desktop or vice versa, emphasizes adaptable grids and flexible design elements. This fluidity ensures that content is readable, accessible, and engaging, irrespective of the screen size.

- **Anticipating future devices**: The dynamism of responsive design also prepares businesses for the future. As new device categories emerge — be they wearables, large touchscreen displays, or something entirely novel — a robust responsive strategy can adapt, ensuring continued relevance in a changing digital landscape.

In the evolving narrative of the multi-device world, responsive design's role is pivotal. It's no longer just about fitting content into varied screen sizes but also crafting experiences that are device-agnostic. Recognizing responsive design as a two-way street is not just about being adaptive but also about being proactive, ensuring that irrespective of how users access the digital world, they are met with experiences that are intuitive, delightful, and deeply engaging.

Reimagining desktop through the mobile lens

In the heart of the digital revolution, as mobile platforms surged with unprecedented momentum, the impact was not just confined to smaller screens. The evolution brought forth by mobile technology began to reimagine and reshape the larger screens of our desktops. It wasn't just about making content accessible on phones and tablets but also about using the mobile lens to reinvent the desktop experience. The aspects that follow illustrate how the desktop can be revisited and redesigned through the mobile lens:

- **Simplicity taking the front seat**: One of the most notable impacts of mobile design on desktop interfaces is the emphasis on simplicity. Mobile interfaces, due to screen size constraints, often prioritize clarity and straightforwardness. When desktop design was viewed through this lens, cluttered interfaces began to transition into more streamlined, user-centric layouts, enhancing user experience.

- **Gesture-inspired interactions**: The touchscreen nature of most mobile devices introduced users to a variety of gestures such as swiping, pinching, and tapping. Desktop designs, influenced by this paradigm, started incorporating similar gesture-inspired transitions and animations, even in non-touchscreen environments, leading to more dynamic and interactive experiences.

- **Focusing on micro-interactions**: Mobile design is filled with micro-interactions — subtle animations and feedback that inform the user about a task's status or guide them through a process. This concept migrated to the desktop realm, adding depth to interactions and making them more engaging and intuitive.

- **Prioritizing speed and efficiency**: Mobile users are often on the go, leading to designs that prioritize quick load times and instant feedback. This ethos has gradually embedded itself into the desktop design. Developers and designers began to reevaluate and optimize desktop applications and websites to ensure they offer similar speed and efficiency.

- **Mobile-first features becoming standard**: Features that were once exclusively designed for mobile, such as sticky navigation bars and off-canvas menus, started making their way into desktop design. These elements not only provided consistency across platforms but also enhanced desktop usability by borrowing tried and tested mobile-centric features.

- **Adaptive content display**: Mobile design taught us the importance of showing users what they need when they need it. By reimagining the desktop through the mobile lens, content began to be displayed more adaptively, based on user behavior, preferences, and context, enhancing relevance and engagement.

While the mobile revolution was initially perceived as a separate entity, it didn't take long for its ripples to touch the shores of desktop design. By reimagining the desktop through the mobile lens, we've not only created a cohesive user experience across devices but also breathed new life into the larger screens that once dominated our digital interactions. The fusion of these worlds underscores a vital principle: innovation in one domain can spark transformative change in another.

Up next, we'll explore how to overcome and navigate the constraints and difficulties of mobile design.

Overcoming mobile design challenges

In the journey of embracing a mobile-first approach, it's essential to understand that mobile platforms present their own set of unique constraints. While the possibilities of designing for mobile are vast, the limitations are equally real. The following are some of the constraints that need to be addressed to craft compelling, functional, and efficient mobile experiences:

- **Screen size limitations**: Perhaps the most evident constraint in mobile design is the limited screen real estate. Unlike the expansive canvases of desktop monitors, mobile screens are compact, requiring designers to prioritize content meticulously. Every pixel counts and the challenge lies in presenting information in a way that's both comprehensive and uncluttered.

- **Touch navigation**: Mobile devices predominantly use touch-based navigation, a stark departure from the mouse-and-keyboard setup of desktops. This means design elements need to be finger-friendly, with adequate space to prevent mis-taps. It's not just about making buttons larger; it's also about rethinking the entire interaction paradigm.

- **Variable connectivity**: While desktops generally enjoy stable, high-speed internet connections, mobile devices often don't have that luxury. They are used in various environments — from high-speed Wi-Fi zones to areas with spotty connectivity. Designing for mobile requires an awareness of these network inconsistencies, ensuring that apps and websites are optimized for performance across varying connection speeds.

- **Battery life considerations**: Mobile devices run on battery power, and certain design decisions can significantly affect battery consumption. High-resolution images, continuous data fetching, and complex animations can drain battery life quickly. It's a delicate balance to provide a rich experience while also being energy efficient.

- **Diverse device ecosystem**: The mobile world is fragmented. Different devices, screen sizes, operating systems, and versions are in circulation. A design that looks impeccable on one device might break on another. This diversity necessitates designs that are flexible and adaptive.

- **Environmental factors**: Mobile devices are used in a myriad of environments: the bright outdoors, dimly lit rooms, noisy streets, or quiet cafés. Recognizing these environmental factors and designing with them in mind — such as ensuring good readability under direct sunlight or providing audio feedback — can significantly elevate the user experience.

- **Limited processing power**: While the processing power of mobile devices has grown by leaps and bounds, it still lags behind that of desktops. This limitation becomes evident when running heavy applications or processing large amounts of data. Optimization is key, ensuring that mobile designs are not just visually appealing but also technically efficient.

In the quest to create the perfect mobile experience, understanding these constraints is half the battle. By recognizing the unique challenges that mobile design presents and crafting solutions tailored to overcome them, designers and developers can create mobile experiences that are not only beautiful but also seamlessly functional.

Navigational nuances — streamlining user journeys

The essence of effective mobile design lies in mastering the art of navigation. In the world of mobile, every pixel of screen real estate is invaluable. As such, providing users with a clear path to their desired destination, without overwhelming them with options or making them dig too deep, is paramount.

One fundamental principle is the prioritization of content and features. Every element, from tabs to buttons, must be carefully curated and positioned, ensuring users can effortlessly find what they're after. This prioritization results in a focused content hierarchy, effectively minimizing the risk of users feeling lost or overwhelmed.

Yet, it's not just about what's on the screen, but also how users interact with it. Familiar navigation patterns, such as the now ubiquitous hamburger menus or tab bars, serve as comforting signposts for users. Building on these well-known patterns ensures a smoother and more intuitive user journey.

But mobile navigation isn't solely about tapping; it's also about swiping, pinching, and dragging. The tactile nature of touchscreens offers an opportunity to incorporate gesture-based interactions. When executed well, these gestures can simplify navigation, but they must feel instinctual rather than forced.

Feedback, too, plays a crucial role. Every interaction, be it a swipe or a tap, should provide some form of acknowledgment. This could manifest as an animation, a color shift, or even haptic feedback. Such cues reinforce the user's actions and guide their next steps.

Another tool in the mobile designer's arsenal is progressive disclosure. Instead of bombarding users with information, provide them with the essentials and offer avenues to explore further. This approach ensures users aren't overwhelmed while still giving them control over their journey.

Lastly, the depth of navigational structures needs consideration. Navigational hierarchies that bury content layers deep can be a source of user frustration. It's always preferable to allow users to reach their goals in as few steps as possible.

To truly excel in mobile design, one must view navigation as more than just a utility. It's an opportunity to guide, engage, and even delight users. Balancing familiarity with innovation, clarity with depth, and guidance with freedom is the challenge — but also the joy — of crafting mobile experiences.

Performance optimization for diverse devices

In an era when smartphones have become an integral extension of ourselves, a seamless user experience across different devices is not just a luxury; it's an expectation. As developers and designers, our role is to ensure that websites and applications perform optimally regardless of the device on which they're accessed. However, achieving this uniformity in performance across the vast landscape of devices is no small feat.

The variance in hardware specifications between devices, be it in processing power, screen resolution, or memory, can drastically affect performance. For instance, an animation that feels smooth on a high-end device might stutter or lag on a budget smartphone. Thus, ensuring a consistent experience demands a holistic approach to design and development, taking into account the lowest common denominator in terms of device capabilities.

A key consideration is image and video optimization. High-resolution visuals, while crisp and appealing, can be resource-intensive. Adapting media based on the device's capabilities and network conditions can drastically cut download times, leading to a smoother user experience. Tools and techniques such as responsive images, adaptive bitrate streaming, and image compression have become invaluable in this endeavor.

The way we structure our code can also make a significant difference. Minimizing and compressing JavaScript and CSS files can help decrease the loading time. Using techniques such as lazy loading, where specific elements are loaded only when needed, can ensure that users are not kept waiting.

Moreover, with the prevalence of varied screen sizes, from the smallest smartphones to larger tablets, ensuring that touch targets are appropriately sized and positioned is critical. An element that is easily clickable on a desktop might be a challenge on a smaller touchscreen.

But it's not just about optimizing for the present. With the rapid evolution of technology, new devices with different capabilities are continuously entering the market. Adopting a future-proof approach, where we design for the devices of tomorrow, ensures longevity in our design solutions.

At the core, the objective remains the same: provide users with a fast, smooth, and consistent experience regardless of the device they choose to use. It's a challenging task, no doubt, but one that lies at the heart of modern mobile-first design.

Ensuring touchscreen responsiveness

Touchscreens have revolutionized the way we interact with digital interfaces. They're intuitive and direct, and offer a sense of immediacy that traditional inputs such as mouse and keyboard can't match. As a developer, ensuring that your application or website is optimized for touchscreen responsiveness is not just a good practice; it's an imperative in today's mobile-centric world.

Imagine the frustration when a user attempts to swipe through a carousel or click a button and there's a noticeable lag or, worse, no response at all. Such experiences can be a quick turn-off and may even deter users from returning to your platform. This underlines the importance of touchscreen responsiveness. It's not just about the action itself, but also the feedback, ensuring that users know their touch has been recognized and is being processed.

Several factors contribute to a responsive touchscreen experience. First, the physical hardware — better quality screens, with higher touch refresh rates, is inherently more responsive. But as developers, the factors we can control revolve around the software and design elements.

One fundamental consideration is touch target size. Elements designed for touch, be they buttons, links, or sliders, need to be of adequate size to accommodate the average finger tap. This prevents missed taps or unintentional actions, which can be a source of user frustration.

Similarly, the feedback mechanism is paramount. Visual cues such as button color changes when tapped or tactile feedback such as subtle vibrations can let the user know their action has been registered. For more complex actions, such as loading a new page, transition animations can serve as both a feedback mechanism and also a delightful design detail.

Also crucial is software optimization. Heavy scripts or unoptimized images can cause lags, rendering even the most meticulously designed touch interface useless. Streamlined code, optimized for performance, ensures that the user's touch is translated into the desired action without any perceivable delay.

Lastly, testing is vital. Different devices, with varied screen sizes and resolutions, can affect touchscreen responsiveness. Regularly testing your design across a spectrum of devices can help identify and rectify any responsiveness issues.

In conclusion, as touchscreens continue to be the primary mode of interaction for a vast majority of users, ensuring a responsive and seamless touch experience is crucial. With meticulous design, thorough testing, and user feedback, developers can create touch interfaces that are both delightful and efficient.

Maintaining aesthetic appeal amid constraints

In the dynamic landscape of mobile design, we're continually walking the tightrope between functionality and aesthetics. Mobile devices, with their unique set of constraints — from screen size to bandwidth limitations — can make it tempting to strip back design elements for the sake of performance. But even within these limitations, the potential to create visually stunning and functional interfaces is immense; it just requires a nuanced approach.

At the heart of this challenge is understanding the essence of mobile use. Mobile users are often on the go, looking for quick answers or short entertainment bursts. They don't have the patience for excessive load times or convoluted navigation. But this doesn't mean they aren't appreciative of a well-thought-out and aesthetically pleasing interface. In fact, a beautiful design can significantly enhance user satisfaction, making them more likely to engage and return.

So, how do you maintain a rich aesthetic appeal amid the constraints of mobile devices?

One approach is embracing minimalism. This doesn't mean sacrificing design elements but rather focusing on the essentials. Streamline your content and design elements to what's absolutely necessary, and then polish those elements to perfection. A minimalist design can reduce clutter, making it easier for users to navigate and find what they're looking for. Additionally, fewer elements can mean faster load times, which is especially crucial for mobile users.

Another technique is using adaptive or responsive imagery. High-resolution images, while beautiful on large screens, can become a burden on mobile devices, leading to longer load times and a subpar user experience. Using adaptive images ensures that the user gets an image that is optimized for their device, maintaining visual appeal without compromising performance.

Typography is another area where aesthetic appeal can shine, even within constraints. Mobile screens, especially on smartphones, have limited space, but that doesn't mean typography should be dull. Embracing bold, readable fonts and playing with hierarchies can ensure that even small screens convey the desired mood and branding.

Lastly, the use of micro-interactions can significantly elevate the user experience on mobile. These are small, often subtle animations or design touches that respond to user activity. A button that changes color when tapped or a subtle transition animation can bring a design to life, making the mobile experience feel more interactive and engaging.

In the realm of mobile design, constraints shouldn't be seen as roadblocks but rather as challenges to be creatively overcome. By focusing on the essentials, optimizing for performance, and paying attention to small details, developers and designers can create mobile interfaces that are both functional and aesthetically captivating.

Summary

As we conclude *Chapter 2*, let's revisit the key lessons and skills we've covered in our exploration of a mobile-first strategy in a multi-device world. This chapter has equipped you with a comprehensive understanding of the critical role that mobile devices play in today's digital landscape and the significant ROI a mobile-centric approach can bring.

We began by examining the dominance of mobile usage, highlighting how it has become the forefront of digital interactions. Understanding this shift is crucial, and you've learned to appreciate the nuances of user behavior in a mobile-first era. This knowledge is vital in adapting web strategies to align with user expectations and trends.

We then delved into the ROI of prioritizing mobile, where you learned about its direct benefits such as enhanced user engagement, better conversion rates, and overall improved user experience. These aspects underscore the importance of a mobile-first approach in achieving business objectives and staying competitive in the digital market.

Additionally, we discussed how optimizing for mobile positively influences other platforms. This holistic view of web strategy ensures that your approach is inclusive and effective across all devices, enhancing your website's overall performance and user satisfaction.

Addressing typical pitfalls in mobile design, we equipped you with the skills to navigate challenges in responsive design, performance optimization, and usability. These practical insights are crucial for overcoming common hurdles in mobile web development.

The lessons from this chapter are invaluable in today's multi-device world. They provide you with a foundation to create web experiences that are not only mobile-friendly but also robust and effective across all platforms.

In the next chapter, we will build upon these foundations. We'll delve into how to create consistent, scalable design systems that ensure seamless experiences across various devices. Understanding design systems is the next natural step, as it brings structure and coherence to the multi-platform strategies we've discussed. Get ready to deepen your skills in creating responsive, user-friendly, and aesthetically pleasing web environments.

3

Design Systems and Responsiveness

Welcome to *Chapter 3*, where we dive into the intricacies and importance of contemporary design systems in the context of **responsive web design** (**RWD**). This chapter is dedicated to deepening your understanding of how meticulously crafted design systems can revolutionize the way web interfaces are developed and experienced across a multitude of devices.

The digital landscape is ever-evolving, and with it, the demands for more sophisticated, scalable, and cohesive web design solutions. In this chapter, we start by exploring the core components of contemporary design systems. You will learn what constitutes a design system, including its typography, color schemes, grid systems, and interactive elements, and why they are pivotal in creating harmonious digital experiences.

Understanding the symbiotic relationship between RWD and design systems is crucial. We will delve into how these systems provide a framework for flexibility and adaptability, ensuring that web designs fluidly respond to different screen sizes, platforms, and user environments. This section aims to equip you with the skills to leverage design systems in enhancing the responsiveness of your web projects.

Next, we focus on the long-term benefits of implementing scalable designs. Scalable design systems are not just about immediate efficiency; they're about setting the groundwork for sustainable, future-proof digital products. You'll comprehend how scalable designs can save time and resources in the long run, facilitate brand consistency, and improve overall user experience.

Finally, this chapter will guide you through implementing consistency in designs across varied devices. Consistency is key to user experience and brand recognition, and you will learn practical ways to maintain this consistency while catering to diverse device specifications and user preferences.

By the end of this chapter, you will have a comprehensive understanding of design systems and their integral role in RWD. You will be equipped with the knowledge and tools to create cohesive, scalable, and responsive digital experiences that meet the demands of today's diverse web landscape. Let's embark on this journey to master the art of creating unified and responsive design systems.

The topics covered in this chapter are the following:

- Defining modern design systems
- Synergy between design systems and RWD
- Strategic benefits of scalable design
- Ensuring design consistency across devices

Defining modern design systems

To appreciate the role of modern design systems in creating responsive, cohesive digital experiences, it's essential to understand their evolution. The concept of a design system is not entirely new; however, its application and significance in the digital world have undergone a transformative journey.

In the early days of digital design, the focus was primarily on individual elements — creating unique buttons, icons, and interfaces for each project. While this allowed for creativity, it often led to inconsistencies, especially as digital products grew more complex. As brands expanded their digital footprint across multiple platforms, the need for a more standardized, systematic approach became evident.

The advent of large-scale digital experiences necessitated a shift. Businesses and design teams recognized the inefficiency of reinventing design elements for each new project. The solution lay in creating a unified system — a library of reusable components, governed by a set of standards and guidelines. This was the genesis of modern design systems.

Design systems, as we know them today, are comprehensive frameworks that combine typography, color schemes, layouts, and interactive elements, all harmonized under a set of standards. They emerged from the need to maintain consistency and quality across various digital touchpoints of a brand. More than just a collection of assets, these systems are underpinned by a philosophy that values coherence, usability, and efficiency.

The proliferation of mobile devices and the rise of RWD further propelled the evolution of design systems. Designers faced the challenge of creating experiences that were consistent across a myriad of screen sizes and platforms. Design systems offered a solution — a central repository of design elements that could be adapted and reused, ensuring uniformity and responsiveness regardless of the platform.

Today, design systems are integral to the strategies of leading digital brands. They enable design and development teams to work more collaboratively and efficiently, reducing the time and resources spent on creating and maintaining digital products. By streamlining the design process, they allow teams to focus more on innovation and user experience rather than getting entangled in the minutiae of design consistency.

In summary, the evolution of design systems reflects the maturing of digital design as a discipline. From disjointed, project-specific efforts to cohesive, scalable frameworks, design systems have become the backbone of modern digital design methodologies. As we delve deeper into the components and applications of these systems, their value in crafting responsive, user-centric digital experiences becomes increasingly apparent.

Components of a design system

Understanding the components that constitute a design system is crucial in grasping its functionality and importance in digital design. A design system is more than a mere collection of reusable assets; it's a comprehensive framework that guides the creation of consistent and scalable digital products. Let's explore the essential components that make up a modern design system:

- **Visual language**: At the core of any design system is its visual language, which includes color palettes, typography, iconography, and imagery. This component ensures visual consistency across all digital interfaces, aligning them with the brand's identity. A well-defined visual language not only aids in brand recognition but also enhances the user's visual experience.

- **UI components**: These are the building blocks of digital products. UI components encompass buttons, input fields, navigation menus, modals, and more. In a design system, these elements are not only standardized in appearance but also in behavior and functionality, ensuring a uniform user experience.

- **Layout and grid structures**: To maintain structural consistency, design systems define grid systems and layout guidelines. These structures provide a framework for arranging UI components harmoniously and responsively across different screen sizes, ensuring a coherent spatial organization.

- **Interactive elements**: Beyond static elements, interactive components such as animations, transitions, and micro-interactions are integral to modern design systems. These elements play a key role in enhancing user engagement and providing feedback during interactions.

- **Design principles and guidelines**: A design system also includes a set of guiding principles and best practices. These guidelines offer directions on how to effectively use the system's components, ensuring that designers and developers adhere to the same standards and approaches.

- **Documentation and standards**: Comprehensive documentation is the backbone of a design system. It provides clear instructions on how and when to use various components, ensuring everyone involved in the product development process has a common understanding.

- **Code libraries and repositories**: For design systems in digital product development, having code libraries and repositories is essential. These resources ensure that the implementation of design components is consistent, efficient, and aligned with established visual and functional standards.

Each component of a design system interplays with the others to create a harmonious and integrated whole. The strength of a design system lies in its comprehensiveness and its ability to guide the creation of diverse yet consistent digital experiences. By standardizing these components, a design system not only streamlines the design and development process but also ensures that the end products are cohesive, responsive, and aligned with the overarching brand strategy.

In the next section, we will delve into how these meticulously crafted components contribute to building and sustaining a brand's presence in the digital world, ensuring that every interaction a user has with the brand is not only consistent but also memorable and aligned with the brand's core values.

Design systems and digital branding

In the digital age, a brand's identity is significantly shaped by its online presence. A design system plays a pivotal role in ensuring this digital branding is not only consistent but also impactful and reflective of the brand's values and personality. Let's examine how design systems contribute to and enhance digital branding.

A brand's digital identity extends beyond its logo or tagline. It encompasses every interaction a user has with the brand online, from the color scheme to the tone of the content, and even the feel of interactive elements. A design system ensures that these elements are not only consistent across various digital platforms but also resonate with the brand's ethos.

The color palette, for instance, is a powerful tool for conveying brand identity. Consistent use of colors across all digital mediums reinforces brand recognition. Similarly, typography in a design system isn't just about legibility; it also communicates the brand's voice — be it professional, friendly, or avant-garde.

Moreover, well-defined UI components in a design system — such as buttons, forms, and navigation menus — become the face of interactions with users. These elements, when consistently and thoughtfully designed, can significantly enhance user experience, creating a sense of familiarity and reliability associated with the brand.

Interactive elements, too, play a vital role. The way elements respond to user interactions — through animations, feedback, or transitions — can leave a lasting impression, subtly reinforcing the brand's character. Whether it's a playful bounce of a button or a sleek slide of a menu, these details contribute to the overall brand experience.

A design system also upholds the brand's identity through its guidelines and principles. By setting standards on how components should be used, the design system becomes a guardian of the brand's digital language, ensuring that every digital product or update aligns with the brand's narrative.

In essence, a design system acts as a bridge between brand identity and user experience. It ensures that every digital touchpoint is an opportunity to reinforce the brand's presence and values. As we delve further into the strategic implications of design systems, we will explore how these systems not only unify and elevate digital branding but also streamline the design process, fostering collaboration and increasing efficiency across teams.

Collaboration and efficiency in design systems

In today's diverse digital ecosystem, where users access content across a myriad of devices, maintaining design consistency is a significant challenge. A well-implemented design system is pivotal in addressing this challenge. It ensures that no matter the device — be it a high-end desktop, a compact smartphone, or a tablet — the design elements and the overall user experience remain consistent.

Design consistency across devices is vital for several reasons. It reinforces brand identity, enhances user experience, and builds trust. When users switch between devices, they expect a seamless transition. Any discrepancies in design can lead to confusion, frustration, and a diminished perception of the brand.

A design system addresses these challenges by providing a unified set of design rules and components. These components are adaptable and responsive, designed to look good and function effectively across different screen sizes and resolutions. For instance, a button that appears on a mobile app should have the same visual style and behavior when accessed on a desktop browser. This consistency in UI elements reinforces brand recognition and makes the user experience more intuitive.

Moreover, a design system includes guidelines on layout and grid structures, ensuring that content is organized and displayed optimally on various screens. This structural consistency is crucial in responsive design, where layouts need to adapt to different screen sizes without losing their coherence or navigational simplicity.

Interactive elements within a design system also play a crucial role. Consistent interactions, such as the response of elements to user input, are key to a cohesive cross-device experience. For example, the way a menu expands with a click or a swipe should be similar, whether on a touchscreen or using a mouse.

In essence, a design system acts as a unifying force in the diverse world of devices, ensuring that the core design principles and brand identity are reflected consistently, no matter how or where the user is accessing the digital product.

As we transition to the next topic, we'll explore how these design systems not only aid in achieving consistency across devices but also how they integrate seamlessly with RWD principles, further enhancing the adaptability and user-friendliness of digital products in a multi-device world.

Synergy between design systems and RWD

The synergy between RWD and design systems is rooted in their shared objective: to create digital experiences that are both visually coherent and functionally robust across a multitude of devices. This foundational synergy is anchored in aligning the core principles of each discipline, leading to a harmonious integration that enhances the overall design and development process.

At its core, RWD is about ensuring web content adapts gracefully to various screen sizes, ensuring usability and aesthetic appeal are maintained regardless of device. It hinges on flexible layouts, responsive images, and media queries to accommodate different viewing contexts. Meanwhile, design systems provide a comprehensive framework of design guidelines, patterns, and components, offering a consistent design language and toolkit for developers and designers.

When these two realms come together, they create a powerful combination. The adaptability of RWD complements the consistency offered by design systems. For instance, the flexible grid layouts advocated in RWD find a natural ally in the modular components of a design system. Together, they ensure that elements not only resize and reflow to fit different screens but also maintain design integrity and cohesiveness.

Moreover, this synergy extends to a philosophical level. Both RWD and design systems are user-centered at their core. They prioritize the user's experience and accessibility, ensuring that digital products are not only accessible to a wider audience but also provide a seamless and engaging user journey. This common ground fosters a unified approach to web design, where every decision is made with the end user in mind.

The union of RWD and design systems also supports the principle of iterative improvement. As user needs and technologies evolve, both frameworks are flexible enough to adapt and grow. This dynamic nature ensures that digital products remain relevant and effective in an ever-changing digital landscape.

In essence, the foundational synergy between RWD and design systems lies in their shared commitment to creating adaptable, user-focused, and consistent digital experiences. This alignment sets the stage for a more streamlined and efficient design process, a topic we will explore in greater depth in the next section. There, we'll delve into how this synergy not only benefits the end product but also enhances the way designers and developers collaborate and execute projects.

Streamlining design processes with unified approaches

The integration of RWD and design systems brings a significant advantage in streamlining design processes. By unifying these approaches, organizations can optimize workflow efficiency, reduce redundancy, and foster a collaborative environment among design and development teams.

Unified approaches in design and development start with a shared understanding and utilization of a common set of tools and guidelines. Design systems offer a set of predefined components and patterns that can be repeatedly used and adapted, significantly reducing the time and resources spent on reinventing designs for each new project. When these systems incorporate RWD principles, these components automatically adapt to various screen sizes and devices, streamlining the development process further.

This amalgamation also enables faster prototyping and iteration. Designers can quickly assemble layouts using the flexible, responsive components from the design system, allowing them to visualize how designs will look across different devices early in the design phase. This approach not only accelerates the design process but also ensures consistency in user experience across different platforms.

Another crucial aspect is the ease of communication and collaboration it fosters between designers and developers. When both teams work within the framework of a unified design system, infused with RWD principles, there's a common language and understanding. This alignment reduces misunderstandings and revisions, leading to smoother project progression and faster delivery times.

Furthermore, this unified approach facilitates easier maintenance and scalability. Changes or updates can be made centrally within the design system and automatically reflected across all implementations. This capability is vital in a digital landscape where continuous improvement and quick adaptability to market changes are essential.

In summary, the synergy between RWD and design systems in streamlining design processes is evident. It leads to more efficient project flows, enhances team collaboration, and ensures consistency and scalability in digital products. Next, we'll delve into how this synergy directly benefits the end users by providing seamless, cohesive, and engaging digital experiences across various devices and platforms.

Enhancing user experience through coordinated efforts

The confluence of RWD and design systems isn't just a technical or procedural improvement; it's a strategic approach that significantly enhances user experience. This synergy, born out of coordinated efforts between these two disciplines, ensures that the end product is not just visually appealing and consistent across devices but also highly functional and user-centric.

Central to this enhanced user experience is the concept of seamless interaction. Whether a user accesses a digital product on a smartphone, a tablet, or a desktop computer, the experience should feel fluid and intuitive. RWD ensures that layouts, images, and interfaces adapt perfectly to different screen sizes, while design systems provide a consistent look and feel across these varied platforms. This consistency in design and functionality reduces user confusion and learning time, making the digital product more approachable and user-friendly.

Moreover, coordinated efforts in RWD and design systems place a strong emphasis on accessibility. By adhering to universal design principles and ensuring that content is accessible across all devices and screen sizes, businesses can cater to a wider audience. This inclusivity not only broadens the user base but also reinforces the brand's commitment to providing equitable user experiences.

The combination of RWD and design systems also allows for more personalized user experiences. Leveraging responsive components that can adapt to not just device types but also user preferences and behaviors, businesses can deliver content and interactions that resonate more deeply with individual users. This level of personalization makes users feel valued and understood, fostering a deeper connection with the digital product.

In essence, the synergy between RWD and design systems in enhancing user experience is profound. It elevates the standard of digital products, making them more accessible, intuitive, and engaging. As we continue to explore the intricacies of this relationship, the next section will delve into how this synergy not only benefits users but also provides strategic advantages to businesses. We'll examine how scalability, one of the core benefits of this integration, plays a crucial role in meeting business objectives and driving digital success.

Strategic benefits of scalable design

In the rapidly evolving world of digital design, scalability has become a non-negotiable aspect for businesses striving for longevity and relevance. Scalability in digital design refers to the ability to grow and adapt digital products to meet changing business needs, market demands, and user expectations efficiently and effectively. This concept is especially crucial as businesses navigate an increasingly competitive digital landscape.

Scalability addresses several critical needs in digital design. Firstly, it allows for the accommodation of a growing user base without losing performance or quality. As more users engage with a digital platform, a scalable design ensures the interface remains fluid, responsive, and capable of handling increased interactions. This adaptability is paramount in maintaining a positive user experience and, in turn, user retention.

Secondly, scalability is vital in supporting the expansion of a business's digital offerings. As a business grows, its digital assets must evolve to include new features, services, or content. A scalable design system enables this expansion without necessitating a complete overhaul of the existing digital infrastructure. This adaptability not only saves time and resources but also ensures a consistent user experience across the evolving digital landscape.

Moreover, the dynamic nature of technology and user trends requires digital products to be flexible and adaptable. Scalability in design means being able to quickly and efficiently integrate new technologies or adapt to emerging user behaviors. This agility ensures that a business's digital presence remains cutting-edge and relevant.

The strategic importance of scalability in digital design can't be overstated. It's a forward-thinking approach, recognizing the need for digital products to be designed not just for the present but with an eye on the future. Scalable design systems are thus a foundational element for any business seeking to establish a robust and enduring digital presence.

As we move to the next section, we'll explore how scalability goes beyond just future-proofing digital assets. It's also a key factor in optimizing the financial investment in digital infrastructure, ensuring that businesses can maximize their ROI over time, while continuously adapting and growing in the digital realm.

Cost-effectiveness of scalable design systems

In today's digital-first business landscape, the cost-effectiveness of design systems cannot be overlooked. Scalable design systems represent a strategic investment that yields long-term financial benefits for businesses. By enabling a more efficient design and development process, these systems significantly reduce both upfront and ongoing costs, making them a financially savvy choice for organizations of all sizes.

One of the primary cost-saving aspects of scalable design systems lies in their reusability. With a comprehensive library of design components and patterns, designers and developers no longer need to create new elements from scratch for each project. This reusability not only cuts down on the hours spent in design and development but also accelerates the entire process, reducing the **time-to-market** (**TTM**) for digital products. The cumulative savings in time and resources can be substantial, especially for businesses that manage a variety of digital products or require frequent updates to their platforms.

Furthermore, scalable design systems contribute to cost efficiency by reducing the need for extensive redesigns or updates. As digital platforms grow and evolve, scalable design systems offer the flexibility to expand and adapt without requiring complete overhauls. This adaptability ensures that businesses can keep their platforms up to date with the latest design trends and user expectations without incurring the significant costs typically associated with major redesigns.

Another aspect where scalable design systems offer cost benefits is in the realm of cross-team communication. By providing a common language and standardized guidelines, these systems streamline collaboration between different teams — such as design, development, and marketing. This enhanced communication reduces misunderstandings and iterations, further saving time and money.

Moreover, the consistency ensured by scalable design systems plays a crucial role in maintaining brand integrity across multiple platforms. The expense of rectifying brand inconsistencies across various digital assets can be significant. Scalable design systems preemptively address this issue, ensuring brand consistency is maintained from the outset, thus avoiding potentially costly corrections down the line.

In summary, the cost-effectiveness of scalable design systems is multifaceted, encompassing savings in design and development time, reducing the need for major redesigns, enhancing cross-team efficiency, and maintaining brand consistency. As we move to the next section, we will delve into how these design systems not only save costs but also empower businesses to respond swiftly to market changes and user feedback, a critical capability in today's fast-paced digital environment.

Rapid deployment and market responsiveness

In the ever-evolving digital market, the ability to quickly adapt and respond to changes is crucial for maintaining a competitive edge. Scalable design systems are instrumental in achieving this agility, enabling rapid deployment of updates and responsiveness to market trends and user feedback.

One of the key advantages of a scalable design system is its facilitation of swift updates and iterations. With a library of pre-designed components and patterns at their disposal, design, and development teams can implement changes or add new features much more quickly than if they were creating them from scratch. This efficiency is particularly beneficial in an environment where user needs and market demands can shift rapidly. Being able to respond to these changes promptly not only keeps a digital product relevant but also demonstrates a commitment to meeting user needs.

Moreover, scalable design systems allow for a more dynamic approach to A/B testing and user feedback incorporation. Teams can test out different design elements or features, gather user input, and then quickly iterate based on this feedback. This process of continuous improvement is crucial for creating user-centric digital products that resonate with the target audience.

Another aspect where scalable design systems aid in rapid deployment is through their ability to maintain consistency across various platforms and devices. When a new feature or design update is rolled out, it can be implemented uniformly across all touchpoints, ensuring a cohesive user experience. This consistency not only streamlines the deployment process but also helps in maintaining brand integrity across the digital landscape.

Additionally, the modular nature of design systems allows for scalability in the true sense — the ability to grow and expand digital products as the business scales. This scalability means that as the market evolves and the business introduces new offerings or enters new markets, the existing digital infrastructure can adapt without the need for extensive redevelopment.

In summary, scalable design systems empower businesses with the agility to quickly deploy updates, respond to market changes, and continuously refine their digital products based on user feedback. This responsiveness is a critical asset in today's fast-paced digital world.

As we turn to the next section, we will explore how scalable design systems not only facilitate immediate responsiveness but also play a pivotal role in ensuring long-term brand consistency. This consistency is key to evolving a brand's digital presence while maintaining its core identity and values, ensuring that the brand remains recognizable and trusted by its users over time.

Long-term brand consistency and evolution

In the digital era, where brand interaction often occurs online, maintaining long-term brand consistency is crucial. This consistency forms the backbone of brand recognition and user trust. The synergy between design systems and RWD plays a vital role in ensuring this consistency, while also allowing for the brand's evolution over time.

Design systems provide a unified collection of design elements and guidelines that encapsulate a brand's visual and interactive language. This consistency is key to building a strong brand identity across various digital platforms. When users interact with a brand, whether on a website, mobile app, or any other digital medium, they encounter a consistent aesthetic and functional experience, reinforcing the brand identity.

The role of RWD in this context is to ensure that the consistency upheld by the design system translates effectively across different device environments. It ensures that the visual and interactive elements of a brand adapt seamlessly to various screen sizes and orientations without losing their inherent character. This adaptability is essential for maintaining brand consistency in a world where users access content on an array of devices.

However, consistency does not imply stagnation. Brands must evolve to stay relevant. Here, the combination of design systems and RWD becomes particularly powerful. Design systems are inherently modular and scalable, allowing for elements to be updated or added without overhauling the entire system. This flexibility enables brands to evolve their digital presence, reflecting changes in their brand strategy or visual identity while maintaining underlying consistency.

For instance, a brand undergoing a rebranding process can update its color scheme, typography, or other design elements within its design system. These updates then propagate across all digital platforms through RWD, ensuring that the evolution is reflected coherently everywhere.

In summary, the synergy between design systems and RWD offers a strategic advantage in maintaining long-term brand consistency and facilitating brand evolution. It ensures that brands present a unified, recognizable face to the world while retaining the flexibility to grow and adapt in an ever-changing digital landscape.

As we transition from understanding the role of design systems and RWD in maintaining brand consistency and facilitating evolution, in the next section, we will delve deeper into how this scalability plays a critical role in business strategy. We will explore how scalable design systems not only contribute to maintaining consistency but also provide operational efficiency and market agility — essential components for any business looking to thrive in the digital age.

Ensuring design consistency across devices

Achieving cross-device consistency is a cornerstone in creating a unified and seamless user experience in digital design. This consistency ensures that users encounter a familiar interface and interaction pattern, whether they are browsing on a smartphone, tablet, or desktop. Let's delve into the fundamental principles that underpin this essential aspect of digital design.

The first principle is the uniformity of core design elements. This includes a consistent use of color schemes, typography, and imagery across all platforms. Such visual coherence strengthens brand identity and aids in user recall. For instance, a user should be able to recognize your brand whether they are engaging with your mobile app or desktop website, with key visual elements remaining consistent.

Another crucial principle is the standardization of navigational structures and UI elements. Users should not have to relearn how to navigate your digital space when switching devices. Elements such as menus, buttons, and links should maintain a consistent design and functionality. This uniformity in navigation and interaction design reduces user frustration and enhances the overall usability of your digital products.

Adaptive content is also key in maintaining cross-device consistency. While the display and layout of content might change across devices, the core message and functionality should remain intact. This approach ensures that users receive the same value and information, regardless of the device they use.

Furthermore, responsive feedback and interaction design play a significant role. The way your application or website responds to user interactions — be it through animations, transitions, or other feedback mechanisms — should be uniform to maintain a coherent experience. Consistent feedback across devices reinforces user actions and enhances their engagement with your product.

Consistency in performance is another principle that cannot be overlooked. Users expect similar responsiveness and speed across different devices. Optimizing performance for various devices, while maintaining high standards, is crucial for a positive user experience.

In conclusion, maintaining cross-device consistency is rooted in these fundamental principles — uniformity in design, standardization of UI elements, adaptive content, consistent feedback, and performance optimization. By adhering to these principles, you ensure that your digital presence is cohesive, user-friendly, and reflective of your brand's identity, regardless of the device used.

As we move to the next topic, we will explore various obstacles and complexities that arise in achieving this consistency across an increasingly diverse landscape of devices and platforms. This understanding is critical in devising strategies to overcome these challenges and maintain a harmonious digital ecosystem.

Challenges in multi-device environments

Navigating the waters of multi-device environments presents a unique set of challenges for designers and developers aiming to maintain design consistency. The diversity of devices in terms of screen sizes, resolutions, and capabilities means that a one-size-fits-all approach is no longer viable. Understanding these challenges is the first step in developing effective strategies to ensure a consistent and cohesive user experience.

Now, let's navigate the challenges:

- **Varied screen sizes and resolutions**: One of the most prominent challenges is the wide range of device screen sizes and resolutions. Designing a layout that looks equally effective on a small smartphone, a tablet, and a large desktop monitor requires careful planning and execution. Elements that work well on a large screen may become impractical or unusable on smaller devices.

- **Diverse user interaction models**: Different devices not only vary in size but also in the way users interact with them. The touch interface of smartphones and tablets offers a different user experience compared to the mouse and keyboard setup of desktops and laptops. Ensuring that navigation and interaction elements are intuitive and efficient across these varying interaction models adds another layer of complexity.

- **Performance across devices**: Performance optimization is a key concern, especially considering the varying hardware capabilities of devices. High-end devices can handle resource-intensive designs with ease, but the same designs may lead to slow performance on older or less powerful devices. Balancing aesthetic appeal with functionality and performance is crucial.

- **Consistent branding in diverse contexts**: Maintaining consistent branding and visual identity across different devices is challenging yet vital. The brand's essence needs to be conveyed effectively, irrespective of the device's limitations or capabilities.

- **Testing and quality assurance**: Rigorous testing across a range of devices is essential to ensure consistency. This process can be time-consuming and resource-intensive but is critical for identifying and rectifying inconsistencies in the design.

- **Adapting to new and emerging devices**: The technology landscape is constantly evolving, with new devices and form factors emerging regularly. Keeping up with these changes and ensuring that designs remain consistent across both current and future devices is an ongoing challenge.

In summary, the multi-device environment presents a complex array of challenges that require thoughtful solutions and strategies. Each of these challenges must be carefully addressed to ensure that users receive a seamless and consistent experience, regardless of their choice of device.

As we transition to the next section, we will explore how RWD serves as a key tool in addressing these challenges. RWD provides a framework for creating designs that adapt fluidly to various device environments, ensuring consistency and enhancing the overall user experience.

Role of responsive design in consistency

RWD is not just a technical solution; it's a strategic approach pivotal for ensuring design consistency across a multitude of devices. RWD plays a crucial role in addressing the challenges of multi-device environments by allowing web content to fluidly adapt to different screen sizes, resolutions, and interaction models. This adaptability is key to maintaining a uniform and cohesive user experience, which is essential for brand consistency and user satisfaction.

Let's explore the core reasons to implement RWD:

- **Fluid grids and flexible layouts**: At the heart of RWD lies the concept of fluid grids. Unlike fixed-width layouts that remain the same regardless of screen size, fluid grids scale dynamically to fit the viewing area. This flexibility ensures that the layout of web content remains consistent and coherent, whether it's viewed on a large desktop monitor or a small smartphone screen.

- **Responsive media and content**: RWD also involves responsive media, including images and videos, that resize and adjust to fit different screen sizes. This ensures that visual elements contribute effectively to the design, regardless of device constraints. Similarly, responsive typography adjusts for readability and impact across devices, maintaining the textual style and tone essential for brand communication.

- **Adaptive UI elements**: RWD allows UI elements such as buttons, menus, and forms to adapt based on the device. This adaptation is crucial for usability, ensuring that interactive elements are not just visually consistent but also functionally consistent. It ensures that users have a familiar interaction experience across devices.

- **Meeting diverse user expectations**: In today's digital landscape, users expect seamless experiences across all their devices. RWD meets this expectation by ensuring that users receive a consistent experience, enhancing usability, and reducing the learning curve associated with switching between devices.

- **Optimizing performance across devices**: RWD also involves optimizing website performance across devices, ensuring that load times and interactivity are consistent. This is crucial for user retention and satisfaction, as performance is a key component of the overall user experience.

In essence, RWD is an indispensable tool in the quest for cross-device design consistency. It enables designers and developers to create experiences that are not only visually coherent across devices but also functionally optimized for various user contexts.

Testing and quality assurance for uniformity

Testing and quality assurance are critical components in the pursuit of ensuring design consistency across various devices. These processes are what ultimately guarantee that the principles of RWD and design systems are executed accurately, offering every user a uniform experience regardless of their choice of device.

Here's a list of the various types of testing:

- **Comprehensive testing across devices**: The first step in this process is extensive testing across a wide range of devices. This includes not only different sizes and resolutions but also varying operating systems and browsers. Such thorough testing ensures that design elements such as layouts, images, and interactive features perform consistently and as intended in diverse environments.

- **Emulating real-user scenarios**: Quality assurance goes beyond technical correctness; it involves emulating real-user scenarios to understand how design elements are experienced in practical use. This approach helps in identifying any usability issues that might not be apparent in a standard testing environment. It's about ensuring that the design feels intuitive and accessible to users with varying levels of technical proficiency and in various usage contexts.

- **Performance testing**: Consistent user experience is not just about visual and functional uniformity; it also includes performance. Load times, response to user interactions, and the smoothness of animations are tested to ensure they meet standards across all devices. Performance testing is crucial, as slow or unresponsive elements can detract significantly from the user experience.

- **Accessibility testing**: An often-overlooked aspect of quality assurance is testing for accessibility. This ensures that the design is usable and accessible to people with disabilities, such as those who rely on screen readers or have limited motor skills. Accessibility is a key component of design uniformity and inclusivity.

- **Iterative testing and feedback loop**: Quality assurance in design is not a one-time process but an ongoing one. It involves setting up iterative testing and feedback loops to continually improve the design. As new devices enter the market and user preferences evolve, regular testing and updates ensure that the design remains consistent and relevant.

In conclusion, testing and quality assurance are the linchpins in ensuring design consistency across devices. They validate the effectiveness of design systems and RWD, making certain that users receive a uniform and high-quality experience, irrespective of how they access digital content.

Summary

As we wrap up this chapter, let's reflect on the key insights and skills we have gained. This chapter has provided a comprehensive exploration of contemporary design systems and their critical role in RWD, equipping you with essential knowledge and tools to enhance your web development projects.

We began by dissecting the components of modern design systems and understanding their importance in creating coherent and scalable web interfaces. You learned about the elements that make up these systems, such as typography, color schemes, and interactive components, and how they contribute to a unified design approach.

Understanding the relationship between RWD and design systems was a focal point of this chapter. We delved into how design systems support and enhance responsiveness, ensuring that web designs adapt seamlessly across different devices and screen sizes. This knowledge is crucial for developing web solutions that provide consistent and optimal user experiences regardless of the device used.

We also examined the long-term benefits of scalable designs. You now understand how scalable design systems not only streamline the development process but also ensure consistency and adaptability in the long run. This approach is key to maintaining brand identity and delivering a consistent user experience across various platforms.

Lastly, the chapter focused on implementing design consistency across varied devices. This skill is vital in today's multi-device world, where users expect a uniform experience whether they are on a desktop, tablet, or smartphone.

The lessons and skills from this chapter are invaluable in your journey as a web developer or designer. They empower you to create designs that are not only aesthetically pleasing but also functionally robust and user-friendly across different devices.

In the next chapter, we will build on these foundations and shift our focus to the strategic planning and execution of content and media. You'll learn how to effectively strategize and adapt your content for diverse screen sizes and resolutions, ensuring that your message is conveyed effectively and efficiently. This is a natural progression from design systems and responsiveness, as content is the heart of every web project, and its presentation is key to user engagement and satisfaction. Let's move forward to mastering the art of content strategy in a multi-device world.

Content and Media Strategy for Varied Screens

Welcome to *Chapter 4*, where we'll delve into the dynamic world of content creation and distribution across a myriad of devices and screen sizes. In this digital era, where content consumption patterns vary vastly across devices, mastering the art of flexible yet consistent content delivery is crucial. This chapter is designed to equip you with essential strategies and skills to effectively navigate this multifaceted landscape.

As we progress through this chapter, you will learn how to make your content flexible while maintaining consistency. This skill is pivotal in ensuring that your message remains coherent and impactful, regardless of the device it is viewed on. We will explore techniques to adapt textual, visual, and multimedia content so that it resonates with your audience across all platforms.

Understanding different media types and how they fit with various devices forms a significant part of this chapter. You'll gain insights into selecting the most appropriate media formats and resolutions for different devices while considering factors such as screen size, resolution, and user context. This knowledge is crucial in creating a media strategy that is both efficient and effective.

Another key focus of this chapter is ensuring that your content retains its essence, regardless of screen size. We will tackle the challenges of presenting content on everything from large desktop monitors to compact mobile screens, ensuring that its fundamental message and quality are not lost in translation.

Lastly, we'll dive into the shift toward dynamic content delivery mechanisms. You'll learn about the latest trends and technologies in content delivery, such as adaptive content and responsive design techniques. Understanding these will enable you to deliver a seamless and engaging user experience that's tailored to the unique demands of each device and user.

By the end of this chapter, you will have a comprehensive understanding of creating and executing a versatile content and media strategy that caters to the diverse landscape of devices today. This knowledge will empower you to craft compelling digital experiences that captivate your audience, regardless of how or where they choose to engage with your content. Let's embark on this journey to mastering content strategy in a multi-device world.

The following topics will be covered in this chapter:

- Content adaptability and flow

- Prioritizing media for different screens

- Retaining message integrity across resolutions

- The rise of dynamic content rendering

Content adaptability and flow

In today's digital ecosystem, which is characterized by a vast array of devices with varying screen sizes and capabilities, content fluidity has become a critical aspect of web and media strategy. Content fluidity refers to the ability of digital content to adapt seamlessly across different devices, providing a consistent and engaging user experience. This adaptability is paramount in a multi-device world where users expect to transition smoothly between their smartphones, tablets, and desktops without losing the context or integrity of the content.

The key to content fluidity lies in understanding that different devices offer varied user experiences. What works on a desktop may not translate effectively on a mobile device and vice versa. For instance, a long-form article may be easily navigable on a desktop but could overwhelm a mobile user. Similarly, interactive elements that are engaging on mobile might not have the same impact on larger screens.

Therefore, creating fluid content involves designing for flexibility. This means thinking beyond fixed layouts and static content; it requires an approach where text, images, videos, and interactive elements dynamically adjust not just in size but also in format and presentation. This flexibility ensures that the content not only fits the screen but also resonates with the user's context, be it glancing at a phone or more in-depth engagement on a desktop.

Moreover, content fluidity also involves considering bandwidth and loading times. High-resolution images that load effortlessly on a desktop with a high-speed connection might hinder the user experience on a mobile device with limited connectivity. Thus, fluid content must be optimized for performance, balancing aesthetic quality with practical loading times.

In essence, understanding content fluidity in a multi-device world is about recognizing and responding to the diverse ways users interact with content across devices. It's about ensuring that the content is not just accessible but also relevant and engaging, regardless of how it's accessed.

As we move forward, we will delve into the foundational strategies that drive the creation of responsive and adaptive content. These principles form the backbone of effective content strategies in today's interconnected, multi-device digital landscape.

Principles of a responsive content strategy

In the realm of digital content, a responsive strategy is more than a technical requirement; it's a user-centric approach to content creation and distribution. A responsive content strategy revolves around designing content that not only adapts to different devices in terms of layout and size but also resonates with the varied contexts in which users access this content. Let's explore the key principles that underpin an effective responsive content strategy:

- **Context is king**: The cornerstone of a responsive content strategy is understanding the context in which your audience will consume the content. This includes not just the device they're using, but also their location, time of day, and the intent behind their interaction with your content. Content should be tailored to fit these varying contexts, providing users with the most relevant and engaging experience.

- **Flexibility and fluidity**: Content should be designed to flow seamlessly across different devices. This means moving away from fixed-width designs to fluid layouts that adapt to any screen size. It also involves using media queries to alter content's presentation and ensure it looks great and functions well, whether on a small smartphone screen or a large desktop monitor.

- **Prioritizing content based on device**: Different devices often equate to different user needs and behaviors. For example, mobile users might prefer quick, bite-sized content, while desktop users might engage more with in-depth materials. A responsive content strategy prioritizes and organizes content based on the likely preferences of users on different devices.

- **Performance optimization**: Fast loading times are crucial, especially for mobile users. This means optimizing images, videos, and other media to ensure they don't hinder page speed. Techniques such as lazy loading, where content is loaded only when it's needed, play a crucial role in performance optimization.

- **Scalable and sustainable content management**: As content needs and technologies evolve, your strategy should be scalable and easy to manage. This involves using a **content management system** (**CMS**) that supports responsive designs and allows for easy updates and management of content across different platforms.

Implementing these principles requires a blend of creative and technical skills, ensuring that content not only adapts to different screens but also engages users in the most effective way in their specific context.

Adaptive content — tailoring for context

Adaptive content is a strategy that goes beyond mere responsiveness; it's about tailoring content to the context in which it will be consumed. This approach recognizes that the user's environment, device, preferences, and behavior influence how they interact with content. By adapting to these factors, content becomes not just accessible, but also relevant and engaging to each user.

Here are some reasons to implement adaptive content:

- **Content personalization**: At the heart of adaptive content is personalization. This involves using data and analytics to understand user preferences and behaviors, and then delivering content that aligns with these insights. For instance, a news website might display different stories based on a user's past reading habits or current location.

- **Context-aware content delivery**: Adaptive content responds to the user's current situation. This means considering the device being used, the time of day, or even the user's current task they are trying to complete. For mobile users in a hurry, summarizing key points or using bullet lists can improve content absorption, whereas desktop users might prefer more in-depth exploration.

- **Dynamic content presentation**: Adaptive content also changes in presentation. This might mean altering the layout, changing the size of interactive elements, or even modifying the navigation based on the device and user context. Such dynamic presentation ensures that content is not just seen but is also interacted with effectively.

- **Seamless content integration across platforms**: Ensuring that adaptive content integrates seamlessly across different platforms is key. A user switching from a mobile app to a desktop site should find consistency in content and presentation, even if the format is adapted for each platform.

- **Scalable and sustainable systems**: Implementing adaptive content requires systems that can handle dynamic changes without frequent manual interventions. This means having a robust CMS capable of handling adaptive content strategies.

Adaptive content is about meeting the user where they are, with what they need, in a way that resonates best with their current context. It's an approach that enhances engagement, improves user experience, and ultimately drives better outcomes for digital platforms.

Next, you'll learn how to introduce effective navigation to aid your content.

Navigational strategies for enhanced content flow

Effective navigation is a critical element in the content strategy, particularly when dealing with varied screen sizes and devices. The goal is to guide users through content smoothly and intuitively, regardless of how they access it. Well-thought-out navigational strategies can significantly enhance content flow, making the user's journey both enjoyable and efficient. Let's explore some strategies:

- **Simplifying navigation for smaller screens**: On smaller devices such as smartphones, screen real estate is limited. Simplifying navigation becomes essential. This can involve using hamburger menus, prioritized navigation elements, or collapsible menus that make the best use of the available space while ensuring users can navigate through content with ease.

- **Consistent navigation across devices**: Consistency in navigation across different devices is key to a seamless user experience. Users should be able to switch from mobile to desktop and intuitively know how to navigate the content. Consistency in menu placement, interactive elements, and navigation style plays a crucial role in this.

- **Using progressive disclosure**: **Progressive disclosure** is an approach where information is presented in a way that shows the most critical elements first, with the option to access more detailed content if needed. This strategy keeps the user from being overwhelmed, particularly on devices where space is a premium, and enhances the content flow.

- **Interactive elements for engagement**: Including interactive elements such as dropdowns, sliders, or tabs can make navigating large amounts of content more manageable and engaging. These elements should be designed to work intuitively across all devices, contributing to a dynamic and interactive content experience.

- **Accessibility in navigation**: Ensuring that navigation is accessible to all users, including those with disabilities, is vital. This includes keyboard-friendly navigation, clear and descriptive labels for screen readers, and consideration of color contrast for visibility.

Effective navigational strategies in content flow are not just about making it easier for users to find what they need; they are about enhancing the overall experience, making content discovery an engaging journey in itself. The design needs to be consistent. So, let's explore how we can achieve that.

Maintaining content integrity across platforms

In a multi-platform digital environment, maintaining the integrity of content across various devices is crucial. Content integrity involves ensuring that the core message, quality, and purpose of the content remain consistent, regardless of the platform or device on which it is viewed. This consistency is key to building a strong, trustworthy brand image and providing a reliable user experience.

Let's look at how we can build consistency across the web:

- **Consistent core messaging**: The essence of content integrity lies in delivering a consistent core message across all platforms. Whether a user reads an article on a mobile app, browses a website on a desktop, or interacts with a social media post on a tablet, the underlying message should remain unaltered. This consistency reinforces the brand's voice and ensures that the key message is communicated effectively, regardless of the medium.

- **Adapting content without altering meaning**: While how content is presented may need to be adapted for different platforms — shorter text for mobile, more interactive elements for desktop — it's crucial that these adaptations do not alter the original meaning or intent. The challenge lies in modifying the format and delivery of content to suit different devices while still conveying the same message and maintaining the same quality.

- **Visual consistency**: Alongside textual content, maintaining visual consistency is equally important. This includes using a consistent color scheme, typography, and imagery style across platforms. Visual elements play a significant role in brand recognition and user perception, making their consistency across platforms vital for content integrity.

- **Quality control across devices**: Ensuring that content maintains a high quality across all platforms is a critical aspect of content integrity. This involves regular quality checks and updates to ensure that all content, regardless of where it is displayed, meets the brand's standards. It's important to consider factors such as readability on different screen sizes, image resolution, and interactive element functionality.

- **Responsive and adaptive design**: Utilizing responsive and adaptive design principles helps in maintaining content integrity. These design approaches ensure that content not only fits different screen sizes but also preserves its functionality and aesthetic appeal, providing a cohesive user experience.

Maintaining content integrity across platforms is not just about consistency; it's about respecting the user's context and delivering a seamless brand experience. It requires careful planning, attention to detail, and a deep understanding of how different platforms are used by the audience. By upholding the integrity of content across all platforms, brands can establish a strong, reliable presence in the digital space.

Next, we'll explore the impact that media prioritization has on different screen sizes.

Prioritizing media for different screens

In today's digital landscape, media consumption patterns vary significantly across different devices. Understanding these variations is crucial for effectively prioritizing and tailoring media content to meet the specific needs and preferences of users on each device. This understanding forms the foundation of a strategic approach to media distribution and presentation in a multi-screen world.

Let's explore these critical considerations:

- **Diverse device usage patterns**: Different devices often cater to different user needs and contexts. For example, mobile devices are typically used for quick, on-the-go interactions, while desktops are more likely associated with longer, more in-depth content consumption sessions. Tablets often fall somewhere in between, offering a more leisurely browsing experience than smartphones but more mobility than desktops. Recognizing these usage patterns is key to determining what type of media content is most appropriate for each device.

- **Screen size and media engagement**: The size of the device's screen significantly impacts how users engage with media. On smaller screens, such as smartphones, users may prefer concise, easily digestible content. Larger screens, such as those of desktops, are more conducive to consuming longer-form videos or detailed images. This correlation between screen size and content type guides the optimization of media for various devices.

- **Bandwidth considerations**: The network connection is another critical factor in media consumption. Mobile users might be on limited data plans or experiencing varying network speeds, which affects their ability to stream high-bandwidth content such as HD videos. On the other hand, desktop users are more likely to have a stable, high-speed connection. Thus, media content needs to be optimized accordingly to ensure accessibility and a good user experience.

- **Interactive media and device capabilities**: The capability of the device also plays a role in media consumption. Advanced interactive media or animations might be well-suited to the powerful processors of desktops but could be overwhelming or underperforming on mobile devices. Understanding these technical limitations and capabilities is essential for delivering an optimal media experience.

- **User preferences and behavior analytics**: Analyzing user behavior and preferences through data analytics can provide valuable insights into how different audiences consume media across various devices. This data-driven approach enables content creators and marketers to tailor their media strategy so that it aligns with user expectations and habits on each device.

Understanding media consumption across devices involves considering multiple factors, including device usage patterns, screen size, network bandwidth, device capabilities, and user preferences. By gaining a deeper understanding of these elements, content creators and marketers can ensure that their media strategy is prioritized and optimized appropriately for different screens, leading to a more engaging and satisfying user experience.

Imagery is one of the most important aspects of the web. So, let's dive into how we can best make use of responsive imagery.

Responsive imagery — techniques and best practices

In a digital environment where content is accessed across a multitude of devices, the importance of responsive imagery cannot be overstated. Responsive imagery refers to images that adapt in size and resolution to suit different devices, ensuring optimal viewing experiences without compromising on load times or visual quality. Implementing responsive imagery effectively involves several key techniques and best practices. Let's take a look:

- **Flexible image sizing**: One of the fundamental techniques in responsive imagery is using fluid image sizes. This means setting image widths and heights in relative units such as percentages rather than fixed units. This approach allows images to scale up or down depending on the screen size, ensuring they fit perfectly within the layout.

- **Image resolution and compression**: High-resolution images are essential for clarity, especially on devices with high pixel densities. However, large image files can significantly slow down page load times. To balance quality and performance, images should be compressed to the smallest possible size without losing their visual appeal. Tools and software that automate this process can be invaluable.

- **Using the srcset attribute**: The `srcset` HTML attribute allows developers to specify multiple image sources for different screen sizes. The browser then automatically selects and displays the most appropriate image based on the screen's resolution and size. This technique is crucial for delivering high-quality images tailored to each device.

- **Art direction with the picture element**: Sometimes, different devices require different versions of an image, perhaps cropped or zoomed differently. The `picture` HTML element, in combination with `source` elements and media queries, offers more control over how images are displayed on various devices, ensuring that the most relevant and impactful part of an image is always visible.

- **Lazy loading for images**: Implementing lazy loading can significantly enhance site performance, especially on pages with a lot of images. With lazy loading, images are only loaded when they are about to enter the viewport, reducing initial page load times and saving bandwidth for users who might not scroll through the entire page.

- **Retina display optimization**: For devices with retina displays, standard images can appear blurry. Serving higher resolution images for these devices, while ensuring that they don't affect performance on standard screens, is a key aspect of responsive imagery.

Incorporating these techniques and best practices in responsive imagery ensures that images are not only visually appealing across all devices but also contribute positively to the overall performance and user experience of the website or application. Responsive imagery is not just about adaptability; it's about creating an optimal balance between aesthetics and efficiency.

Now, let's move on to how to deal with video content.

Video content adaptation for various screens

Adapting video content for various screens is a crucial aspect of modern content strategy, especially in a digital ecosystem where users access media on a diverse range of devices. From high-definition desktop monitors to compact smartphone screens, each device offers a unique viewing experience. Ensuring that video content is optimized for these varied environments is essential for maintaining engagement and providing a seamless user experience.

Let's take a look at various techniques we can use to help with video content:

- **Responsive video embedding**: The foundation of video content adaptation is responsive embedding. This involves using CSS and HTML to ensure that video players resize themselves according to the screen's dimensions. Techniques such as aspect ratio boxes can be employed to maintain the correct size and proportion of videos without them appearing stretched or compressed.

- **Adaptive bitrate streaming**: One of the most effective techniques for video adaptation is adaptive bitrate streaming. This technology dynamically adjusts the quality of a video stream in real-time based on the user's available bandwidth and device capabilities. It ensures that the video plays smoothly on various devices, providing the best possible quality without buffering issues.

- **Consideration for mobile data usage**: When adapting videos for mobile devices, it's important to consider users' data plans. Offering lower bitrate options or enabling users to choose their preferred video quality can enhance the mobile viewing experience while being mindful of data consumption.

- **Touchscreen interactivity for mobile devices**: For mobile devices, the video player's interactivity should be optimized for touch screens. Larger playback controls, easy-to-access volume settings, and swipe gestures for skipping or rewinding can significantly improve usability on smaller screens.

- **Captioning and accessibility**: Ensuring that videos are accessible is crucial. This includes providing captions or subtitles for users who are hearing impaired or for those who might watch videos without sound, a common scenario on mobile devices. Captions should be legible on all screen sizes.

- **Testing across devices**: Rigorous testing across different devices is vital to ensure that videos perform well under various conditions. This includes checking load times, playback smoothness, and quality across a range of devices with different screen sizes and resolutions.

By implementing these strategies for video content adaptation, content creators can ensure that their videos are not only visually appealing but also functional and accessible across different screens. This adaptability is key to engaging a diverse audience and catering to their varying viewing preferences and conditions.

Leveraging adaptive media elements

In the realm of digital content, the use of adaptive media elements is essential for creating a versatile and engaging user experience across various devices. Adaptive media elements are designed to adjust in size, resolution, and functionality based on the user's device, ensuring optimal display and interaction. This adaptability is crucial in a multi-device world where users expect seamless access to content, regardless of how they choose to view it.

Here's a list of some of the most prominent adaptive media elements:

- **Scalable Vector Graphics (SVG)**: SVG is a prime example of adaptive media elements. Unlike raster images (such as JPEGs or PNGs), SVGs are composed of vectors, which means they can scale up or down without losing clarity. This makes them ideal for logos, icons, and other graphic elements that need to maintain sharpness on all screen sizes, from the smallest smartphones to the largest desktop monitors.

- **Responsive icons and buttons**: Icons and buttons are integral to user navigation and interaction. When these elements are adaptive, they not only change size but also scale their functionality to suit different devices. For instance, buttons may increase in size for touchscreens to facilitate easier interaction, or icons may change in design to communicate different actions or states more effectively on smaller screens.

- **Flexible audio and video players**: Media players should adapt to various screen sizes and resolutions while maintaining functionality and user experience. This includes repositioning controls for easier access on mobile devices and ensuring that media players are not overly intrusive or small on different screens. Techniques such as responsive design and CSS media queries can be employed to achieve this flexibility.

- **Adaptive typography**: Typography is a critical element of media strategy. Adaptive typography adjusts not only in size but also in weight and spacing to enhance readability and aesthetic appeal across devices. This approach ensures that text elements are legible and visually harmonious, regardless of the viewing platform.

- **Interactive media elements**: Interactive media, such as infographics or data visualizations, should be adaptable to offer the same level of engagement on all devices. This might mean reconfiguring layouts or altering interaction models (such as hover effects on desktops versus touch interactions on mobile devices).

- **Testing for consistency and performance**: Rigorous testing is key to ensuring that adaptive media elements perform consistently across different devices. This includes checking load times, ensuring media elements are displayed correctly, and verifying that interactive components function as intended.

Incorporating adaptive media elements into a content strategy is not just about ensuring compatibility with various devices; it's about enhancing the overall user experience. By thoughtfully leveraging these elements, content creators and designers can ensure that their media is not only accessible but also engaging and effective, no matter how or where it is accessed.

Now, it's time to think about how to align visuals with performance.

Balancing visual appeal with performance

In the digital landscape, the balance between visual appeal and performance is a critical consideration for media strategy. High-quality, visually stunning media can captivate and engage users, but if it compromises performance, particularly in terms of load times and responsiveness, it can detract from the user experience. Achieving a balance between these two aspects is essential for delivering content that is both appealing and functional across various screens.

Here are the concepts you should consider when balancing visual appeal with performance:

- **Understanding the impact of media on performance**: The first step in balancing visual appeal with performance is recognizing how different media elements affect website or app performance. High-resolution images, videos, and complex graphics can significantly increase page load times, especially on mobile devices with limited bandwidth or processing power.

- **Optimal image selection and compression**: Selecting the right format and size for images is crucial. Formats such as JPEG, PNG, and WebP each have their strengths in different scenarios. Effective compression techniques can substantially reduce file sizes while maintaining image quality. Tools and software that automate image optimization based on the device and context can be invaluable in this process.

- **Prioritizing content and lazy loading**: Implementing lazy loading for images and videos ensures that these elements are only loaded when they are needed — typically when they enter the viewport. This technique can improve initial page load times, reducing the wait time for users and improving the overall user experience.

- **Video optimization for various screens**: For video content, providing different resolution options is key. Lower resolutions can be defaulted for mobile devices to ensure quicker loading, with the option to switch to higher resolutions on faster connections or larger screens. Techniques such as adaptive bitrate streaming can dynamically adjust video quality based on the user's current network conditions.

- **Balancing aesthetics and functionality**: The aesthetic aspect of media shouldn't be compromised entirely for performance. It's about finding creative ways to present visually appealing content without overburdening the performance. This might involve using CSS and JavaScript effectively to create engaging layouts and interactions that are less resource-intensive.

- **Regular performance testing and monitoring**: Continuously monitoring and testing the performance of media-rich websites and applications across various devices is essential. This ongoing process helps in identifying any issues that might arise from new content or updates and ensures that a balance is maintained over time.

Achieving a harmonious balance between visual appeal and performance requires carefully planning, selecting, and optimizing media elements. By considering the impact of each media element on performance and employing strategic optimization techniques, content creators and developers can ensure that their media enhances rather than hinders the user experience.

Now, let's talk about resolutions.

Retaining message integrity across resolutions

In the diverse ecosystem of digital screens and devices, resolution plays a pivotal role in content delivery and user experience. Understanding how different resolutions impact the way content is perceived and interacted with is crucial for ensuring that the core message of your content is communicated effectively across various platforms. Let's dive further into exploring resolutions:

- **The role of resolution in content clarity**: Resolution, which refers to the number of pixels on a screen, directly affects the clarity and sharpness of the content displayed. High-resolution screens, such as those on many modern smartphones and tablets, can display content with incredible detail, making them ideal for high-definition images and videos. However, this also means that any flaws or issues in media quality are more noticeable.

- **Adapting content to screen resolutions**: When content is viewed on screens with different resolutions, it must be adapted accordingly to maintain its effectiveness. For instance, an image that looks crisp on a high-resolution desktop monitor might appear blurred or pixelated on a lower-resolution mobile device if not properly optimized. Therefore, it's essential to create and deliver content in various resolutions to ensure it appears sharp and engaging on all devices.

- **Scalability of visual elements**: Scalable elements such as vector graphics and responsive typography play a significant role in maintaining visual quality across resolutions. Vector graphics remain sharp and clear at any size, making them ideal for logos and icons, while responsive typography adjusts to ensure text is readable and aesthetically pleasing on screens of all resolutions.

- **Impact on user experience and engagement**: The resolution can significantly impact user engagement and experience. High-quality, clear content can capture and hold users' attention, whereas poor-resolution content may lead to a negative user experience and reduced engagement. Users are more likely to trust and engage with content that is professionally and clearly presented.

- **Testing across different resolutions**: To ensure content integrity, it's crucial to test how your content appears across a range of devices with varying resolutions. This testing helps identify any resolution-related issues that could affect content clarity and user experience.

Understanding the impact of resolution on content delivery involves acknowledging that different devices display content differently and adapting your content strategy accordingly. By ensuring that your content is optimized for various resolutions, you can maintain its integrity, effectiveness, and appeal, regardless of how or where it is accessed.

Now, let's explore various techniques that help us deal with varied resolutions.

Designing for clarity — techniques for all resolutions

Designing content that retains clarity across all resolutions is a fundamental aspect of a robust digital strategy. In a world where users access content on an array of devices, from high-resolution desktop monitors to smaller mobile screens, it is essential to employ techniques that ensure content maintains its integrity and effectiveness regardless of the resolution.

Here are the key strategies to achieve this:

- **Embracing vector graphics**: Vector graphics are essential in maintaining visual clarity across different resolutions. Unlike raster images, vectors are not made up of pixels but paths, which allows them to scale up or down without losing quality. Utilizing SVG for logos, icons, and illustrations ensures that these elements remain crisp and clear on any screen.

- **Responsive typography**: Typography must be adaptable to different screen resolutions. This involves choosing legible font styles and employing relative units such as "em" or "rem" for font sizes, ensuring that text scales appropriately. Additionally, line height and letter spacing should be adjusted dynamically to enhance readability across devices.

- **Adaptive image sizing and compression**: For raster images, use responsive techniques that serve different image sizes for different resolutions. Techniques such as *srcset* and sizes attributes in HTML allow browsers to choose the most appropriate image size, balancing quality and load time. Effective compression methods also reduce file size while maintaining visual fidelity.

- **Prioritizing the content hierarchy**: On smaller or lower-resolution screens, prioritizing key information becomes crucial. Employ a content hierarchy that adapts to different screens, ensuring that the most important information is always prominent, and less critical elements are secondary or hidden in collapsible menus.

- **Using media queries for design adaptation**: Media queries in CSS are a powerful tool for adapting your design to different resolutions. They allow you to apply different styling rules based on screen characteristics, ensuring that your layout and content look their best on any device.

- **Testing across a range of devices**: Regular testing on various devices and resolutions is vital to ensure that all your content design techniques are effective. This should involve not just visual inspection but also user testing to gather feedback on the clarity and usability of the content.

By employing these techniques, you can ensure that your content is not only visually appealing but also clear and accessible to all users, regardless of the device or screen resolution they use. This approach to design is essential for maintaining the integrity and effectiveness of your digital content across the diverse spectrum of today's devices.

As we continue to explore how to maintain message integrity across various resolutions, the upcoming section will delve into scalable graphics and text, further emphasizing the importance of quality and adaptability in digital content creation.

Scalable graphics and text — ensuring quality

In the multi-resolution landscape of modern digital devices, scalable graphics and text are essential for ensuring the quality and integrity of content. As users switch between devices, from high-resolution monitors to compact mobile screens, maintaining the visual and textual clarity of content is crucial. Here's how scalable graphics and text can be utilized effectively to ensure quality across various resolutions:

- **Utilizing vector graphics for scalability**: Vector graphics are a cornerstone of resolution-independent design. Unlike raster images, which can become pixelated or blurry when scaled, vector graphics maintain their clarity and sharpness at any size. This is because they are based on mathematical equations rather than pixels. Formats such as SVG are ideal for logos, icons, and other design elements that need to stay sharp on screens of all sizes.

- **Adaptive typography for readability**: Typography must be responsive to ensure readability across different devices. This involves using relative units for font sizes, such as **viewport width** (**vw**) or ems, which allow text to scale concerning the size of the screen or parent element. Additionally, consider the line height, font weight, and letter spacing to ensure that text is legible and comfortable to read on all devices.

- **High-DPI media queries for high-resolution displays**: High-DPI, where **DPI** stands for **Dots Per Inch**, displays, such as Apple's "Retina" screens, require high-resolution images to avoid blur and pixelation. Media queries can be used to detect high-DPI screens and serve higher-quality images accordingly. This ensures that images appear crisp and detailed on these displays.

- **Responsive scaling of images**: Beyond resolution, the physical size of images needs to be flexible. Using CSS techniques, images can be made to scale within their containing elements, ensuring they fit well within different layouts and screen sizes. This responsive scaling is crucial for maintaining a cohesive and visually appealing layout across devices.

- **Testing across devices and resolutions**: Ensuring quality through scalable graphics and text requires thorough testing across a range of devices and resolutions. This not only involves checking for clarity and legibility but also assessing the overall aesthetic appeal and consistency of the content.

- **Balancing performance and quality**: While striving for high-quality graphics and text, it's important to balance these elements with performance considerations. Techniques such as image compression and choosing the right file format can significantly reduce file sizes without compromising quality, ensuring quick load times even on less powerful devices.

Scalable graphics and text are vital in creating digital content that is versatile and effective across the diverse range of devices used today. By focusing on these elements, content creators and designers can ensure that their message remains clear, engaging, and visually appealing, no matter where or how it is viewed.

Let's explore the role of responsive layouts in the message integrity of our designs.

Responsive layouts and their role in message integrity

Responsive layouts play a crucial role in ensuring message integrity across various screen resolutions. As users interact with digital content on an array of devices, from large desktop monitors to compact smartphones, the layout must adapt while keeping the core message and user experience consistent. This adaptability is essential for maintaining the effectiveness and coherence of the content.

Here are the roles of responsive layouts:

- **Fluid grids for flexible layouts**: At the heart of responsive design are fluid grid systems. Unlike fixed grids bound to specific pixel dimensions, fluid grids use relative units such as percentages, which allow elements to resize in proportion to the screen. This flexibility ensures that the layout adjusts to fit any screen size, maintaining the structure and hierarchy of the content.

- **Prioritizing content with media queries**: Media queries in CSS allow for the customization of content presentation based on different screen characteristics. They enable designers to prioritize and rearrange content elements to suit different viewing contexts, ensuring that the most important information remains front and center, regardless of screen size.

- **Maintaining hierarchical and visual consistency**: A key aspect of retaining message integrity is maintaining a consistent visual hierarchy across devices. This involves ensuring that headings, subheadings, and other key textual elements retain their prominence and that visual cues are used consistently to guide users through the content.

- **Adaptive content and conditional loading**: Responsive layouts often employ adaptive content strategies, where content is not just resized but conditionally loaded or adapted based on the device. For instance, additional information or secondary content might be loaded for a desktop view but omitted or collapsed for mobile so that the most critical elements of the message can be focused on.

- **Touch and mouse interactions**: Responsiveness also extends to interaction models. Layouts must account for different interaction modes — touch for mobile and tablet devices, and mouse for desktops. This includes considering the size and spacing of interactive elements to ensure they are easily navigable on all devices.

- **Regular testing and updates**: Ensuring the integrity of the message through responsive layouts requires ongoing testing and updates. This involves not only technical testing across different devices and resolutions but also user testing to gather feedback on the clarity and effectiveness of the message.

Responsive layouts are more than a technical solution; they are a strategic approach to content presentation. By allowing for the flexible and dynamic display of content, responsive layouts ensure that the core message is consistently and effectively communicated across the vast landscape of devices and screen sizes. This consistency is vital for delivering a coherent and engaging user experience, which is crucial in today's multi-device world.

Testing and optimization strategies for diverse resolutions

In an era where digital content is accessed across a spectrum of devices, each with different resolutions, testing and optimizing for these diverse conditions are imperative. This process is crucial to ensure that content not only looks visually appealing but also maintains its intended message and functionality across all platforms. Here are some key strategies and practices for effective testing and optimization across various screen resolutions:

- **Cross-device and cross-browser testing**: Comprehensive testing across a range of devices and browsers is essential. This includes checking content on different operating systems, screen sizes, and browsers to ensure consistent rendering. Tools such as emulators and real-device testing platforms can be invaluable for simulating various environments and identifying any resolution-specific issues.

- **Implementing responsive design testing**: Testing responsive designs involves more than checking if the content fits on various screens. It requires a detailed examination of how content elements such as images, text, and interactive features adapt and reflow. Pay particular attention to critical breakpoints where layout changes occur to ensure that the content transitions smoothly between different screen sizes.

- **Image and media optimization**: High-quality images and media are often necessary for compelling content, but they can be problematic for performance, especially on devices with lower resolutions or slower internet connections. Implement techniques such as image compression, appropriate file format selection (such as WebP or JPEG XR), and responsive image solutions (such as *srcset*) to optimize media for different resolutions.

- **Performance testing for different resolutions**: Analyze how content performs across devices with varying resolutions. Performance testing should include assessing loading times, interactivity responsiveness, and the impact of media elements on overall site speed. Utilize performance tools to identify bottlenecks and optimize accordingly.

- **Accessibility checks**: Ensure that content remains accessible across resolutions. This includes verifying that text remains readable, interactive elements are easily navigable, and color contrasts are maintained. Tools that simulate various visual impairments can be particularly helpful in this regard.

- **Continuous monitoring and iterative improvement**: Digital landscapes and user devices are constantly evolving. Regularly monitoring and updating content is necessary to maintain optimization across all resolutions. Collect and analyze user feedback and usage data to make informed improvements.

Effective testing and optimization strategies for diverse resolutions are pivotal in ensuring that digital content delivers a consistent and high-quality user experience. By rigorously testing and optimizing for different screen sizes and resolutions, content creators and designers can ensure their message remains clear, engaging, and accessible to all users, irrespective of how they access the content.

Next, we'll learn how dynamic content rendering has become the standard in web design.

The rise of dynamic content rendering

Dynamic content rendering is increasingly becoming a cornerstone in modern web design and content strategy, particularly as the variety of devices used to access web content continues to expand. This approach involves creating web content that adapts and changes in real-time based on user interactions, device specifications, and other contextual factors.

Now, to get an even clearer understanding, let's dive into an overview of dynamic content rendering and its significance in the current digital landscape.

At its heart, dynamic content rendering is about delivering content that is not statically fixed but is instead generated based on specific parameters or conditions. This could mean displaying different content for different users, changing content based on the time of day, or adapting the content layout for various screen sizes and resolutions.

The primary aim of dynamic content rendering is to enhance the user experience. By presenting content that is tailored to the user's context, preferences, or actions, websites, and applications can provide a more personalized and engaging experience. This level of personalization can lead to increased user engagement, satisfaction, and ultimately, higher conversion rates.

Various technologies enable dynamic content rendering. On the client side, JavaScript frameworks such as React or Angular can update the **Document Object Model (DOM)** in real-time, changing the content without the need to reload the entire page. On the server side, technologies such as Node. js can dynamically generate HTML based on the user's context before the page is sent to the browser.

Dynamic content rendering presents unique challenges and opportunities for **search engine optimization (SEO)**. While search engines have become better at indexing dynamically generated content, it's crucial to ensure that dynamically rendered pages are accessible and crawlable. Techniques such as server-side rendering or dynamic rendering can help improve SEO while still offering the benefits of dynamic content.

In a world where content is accessed across a multitude of devices, dynamic content rendering allows for the creation of responsive, adaptable content that looks and functions optimally on any device. This adaptability is especially important for ensuring consistent user experiences across desktops, tablets, and smartphones.

Dynamic content rendering represents a shift toward more interactive, responsive, and personalized web experiences. As user expectations continue to evolve toward more tailored content, mastering dynamic rendering becomes increasingly important for web developers and content strategists aiming to create cutting-edge digital experiences.

Let's explore some technologies that make dynamic rendering possible.

Technologies powering dynamic rendering

Dynamic content rendering, which is essential in today's diverse digital landscape, is made possible by a variety of advanced technologies. These technologies enable content to dynamically adapt and change in response to user interactions, device characteristics, and other contextual factors. Understanding these technologies is key to leveraging the power of dynamic rendering effectively. So, let's begin:

- **JavaScript and Asynchronous JavaScript and XML (AJAX)**: JavaScript, particularly when used in conjunction with AJAX, is foundational for dynamic content rendering. AJAX allows web pages to be updated asynchronously by exchanging data with a web server in the background. This means parts of a web page can be updated without reloading the entire page, leading to a smoother and more interactive user experience.

- **Frontend frameworks and libraries**: Modern frontend frameworks and libraries such as React, Angular, and Vue.js have revolutionized the way dynamic content is rendered. These tools provide developers with the means to build complex, interactive web applications that can update in real-time. They utilize a virtual DOM to efficiently update the browser's displayed content in response to user interactions.

- **Server-side technologies**: Server-side technologies such as Node.js and ASP.NET are used to dynamically generate web content on the server before it's sent to the client. This approach is beneficial for SEO and ensures compatibility with browsers that may have limited JavaScript functionality.

- **CMS**: Many contemporary CMS platforms now support dynamic content rendering. They allow content creators to design and manage content that can change dynamically based on predefined rules or user interactions.

- **Application programming interfaces (APIs) and microservices**: APIs play a crucial role in dynamic content rendering, especially in fetching real-time data that can be used to update content dynamically. Microservices architecture further enhances this by allowing different parts of a web application to be developed, deployed, and scaled independently.

- **Progressive web apps (PWAs)**: PWAs use modern web capabilities to deliver app-like experiences to users. They leverage dynamic rendering to provide fast, engaging experiences that work reliably on all devices and network conditions.

The choice of technology often depends on the specific requirements of the project, including factors such as the target audience, type of content, and desired level of interactivity. By harnessing these technologies, developers and content strategists can create highly responsive, engaging, and personalized web experiences that meet the demands of today's varied and dynamic digital landscape.

Personalization and user-centric strategies

In the realm of dynamic content rendering, personalization stands as a key strategy that's central to enhancing user engagement and satisfaction. By tailoring content to meet individual user preferences, behaviors, and contexts, websites and applications can offer a more relevant, engaging, and user-centric experience. Let's explore how dynamic content rendering facilitates personalization and the strategies involved in implementing it effectively:

- **Understanding user preferences and behaviors**: The first step in personalization is gathering and analyzing data about users' preferences, behaviors, and interactions. This data can come from various sources, such as browsing history, user profiles, and engagement metrics. By understanding what content resonates with users, dynamic rendering systems can deliver more relevant and personalized content.

- **Implementing dynamic user profiling**: Dynamic user profiling involves creating a profile of each user that evolves based on their interactions and preferences. This profile then informs the dynamic content rendering system, allowing it to display content that aligns with the individual user's interests and needs.

- **Context-aware content delivery**: Beyond just user preferences, personalization also involves understanding and responding to the user's current context. This includes factors such as the device being used, the time of day, location, and even current weather conditions. Context-aware content delivery ensures that the content is not only personalized but also appropriate for the user's current situation.

- **Real-time adaptation and feedback loops**: Effective personalization requires the content to adapt in real-time. This involves creating feedback loops where the system continuously learns from user interactions and adjusts the content accordingly. For instance, if a user frequently interacts with certain types of articles or products, the system can start displaying more of that content type.

- **Balancing personalization with user privacy**: While personalization can significantly enhance the user experience, it's important to balance this with respect for user privacy. Transparent data practices and giving users control over their data are essential. Adhering to privacy regulations and ethical standards ensures that personalization strategies are respectful and trustworthy.

- **Testing and optimization**: Personalization strategies should be regularly tested and optimized. A/B testing different personalization approaches can provide insights into what works best for different user segments. Continuous optimization ensures that the personalization strategy remains effective and relevant over time.

Personalization, powered by dynamic content rendering, represents a shift toward more user-centric digital experiences. By delivering content that's tailored to individual users and their contexts, businesses can create more meaningful and engaging interactions, fostering a deeper connection with their audience.

Next, we'll need to think about how to best create a performant website that's optimized for SEO.

Optimizing for performance and SEO

Incorporating dynamic content rendering into a website's content and media strategy demands a careful balance between performance and SEO. While dynamic rendering provides a robust platform for personalization and user engagement, it also poses unique challenges for website performance and visibility in search engines. Here's how to address these challenges effectively:

- **Ensure fast load times**: Dynamic content, especially when loaded on the client side, can potentially slow down a website. To mitigate this, optimize backend server performance and use techniques such as lazy loading for content and images. This ensures that only necessary content is loaded initially, improving page load times, which is a crucial factor for both user experience and SEO.

- **Structured data and metadata**: Use structured data to help search engines understand and categorize dynamically rendered content. Properly implemented metadata ensures that even if the content is loaded dynamically, essential information about the page is available to search engines.

- **Keep URLs consistent**: Ensure that dynamic content changes do not affect the URL structure. Consistent URLs are essential for SEO as they allow search engines to index content correctly and users to bookmark and share links. Implement the History API in JavaScript for any content changes that should be reflected in the URL without reloading the page.

- **Mobile optimization**: With the increasing prevalence of mobile devices, optimizing dynamic content for mobile is critical. This includes not only responsive design but also considering mobile load times and interactivity, which are key factors in Google's mobile-first indexing.

- **Continuous monitoring and optimization**: SEO and performance optimization for dynamic content is not a one-time task. Continuously monitoring site performance, user engagement metrics, and search engine rankings is crucial. Regular audits and updates based on these insights help maintain optimal performance and SEO effectiveness.

Balancing the dynamic nature of personalized content with the necessities of SEO and performance optimization requires a strategic approach. By focusing on these key areas, websites can leverage the benefits of dynamic content rendering while ensuring they remain fast, user-friendly, and easily discoverable by search engines.

Now, let's explore some challenges and their solutions in dynamic rendering.

Challenges and their solutions in dynamic rendering

Dynamic content rendering, while offering enhanced user experiences and personalization, brings its own set of challenges. Addressing these effectively is crucial for maintaining the performance, accessibility, and SEO-friendliness of a website. Here, we'll explore common challenges associated with dynamic rendering and practical solutions to overcome them:

- **Performance overhead**: Dynamic rendering can be resource-intensive, potentially slowing down a website. This is particularly true for client-side rendering where the browser handles most of the rendering workload:

  - **Solution**: Optimize code efficiency and leverage **server-side rendering (SSR)** or **static site generation (SSG)** where appropriate. Use caching strategies and **content delivery networks (CDNs)** to reduce load times. Minimize the use of large libraries and frameworks and adopt lazy loading for media and non-critical content.

  - **Solution**: Implement dynamic rendering solutions such as hybrid rendering, where the initial page load is server-side rendered, ensuring that content is crawlable by search engines. Utilize tools such as Google's Dynamic Rendering, which serves a version of the content optimized for search engine bots.

- **Accessibility concerns**: Dynamic content can sometimes pose challenges for users with disabilities, especially if it relies heavily on JavaScript, which may not always be compatible with screen readers and other assistive technologies:

  - **Solution**: Ensure adherence to web accessibility standards. Provide alternative content for critical dynamic elements and ensure that navigation and interactive elements are accessible through keyboard and assistive technologies.

- **Complexity in development and maintenance**: Implementing dynamic rendering can add complexity to the development process, requiring a higher level of expertise and potentially increasing maintenance overhead:

  - **Solution**: Adopt a modular approach to development where components are reusable and easily maintained. Keep documentation updated and invest in training for development teams. Consider using frameworks and tools that abstract some of the complexities of dynamic rendering.

- **Mobile device constraints**: Dynamic rendering, especially client-side, can be demanding for mobile devices, which may have limited processing power and slower internet connections:

  - **Solution**: Optimize content for mobile, focusing on minimalistic design and reduced reliance on heavy scripts. Implement adaptive designs and ensure media files are compressed and optimized for mobile use.

- **User experience inconsistencies**: Ensuring a consistent user experience across various devices and browsers can be challenging with dynamic content:

  - **Solution**: Conduct thorough testing across a range of devices and browsers to identify and fix inconsistencies. Gather user feedback regularly to understand how different audiences interact with the content and make necessary adjustments.

Dynamic content rendering requires a thoughtful approach to overcome its inherent challenges. By implementing these solutions, organizations can harness the power of dynamic rendering to deliver rich, personalized, and engaging content experiences while maintaining robust performance, SEO, and accessibility standards.

Summary

As we conclude this chapter, let's take a moment to reflect on the essential lessons and skills that have been imparted throughout. Our journey has equipped you with crucial strategies for adapting content and media to a wide range of devices, ensuring that your message is communicated effectively, regardless of how your audience accesses it.

We started by mastering the art of making content flexible yet consistent. You learned the importance of maintaining a coherent narrative and visual style across various platforms, ensuring that your content resonates with your audience, whether they're on a desktop, tablet, or smartphone.

Understanding which media types suit which devices was another key focus. We delved into the nuances of selecting appropriate formats and resolutions for different screen sizes, optimizing for both performance and engagement. This skill is invaluable in a landscape where user experiences are defined by the compatibility of content with their device of choice.

Ensuring that content retains its essence regardless of screen size was a pivotal lesson. We explored techniques to modify content presentation without compromising its core message, ensuring clarity and impact on both large and small screens.

Finally, we discussed the shift toward dynamic content delivery mechanisms. You learned about emerging trends and technologies that enable content to adapt in real-time to different user contexts, enhancing personalization and engagement.

These skills and insights are crucial for anyone looking to create impactful digital experiences in today's diverse device landscape. They not only enhance the reach and effectiveness of your content but also ensure that your audience receives a seamless and engaging experience.

In the next chapter, we will build upon these foundations by focusing on how to optimize your web presence for performance. This progression is natural and necessary as the best content strategy can only be successful if it is underpinned by a robust, high-performing website.

5

Optimization and Performance — a Business Imperative

Welcome to *Chapter 5*, where we delve into the critical role of web performance in the success of your online presence. In this digital era, where milliseconds can make a difference in user engagement and business success, understanding and optimizing your website's performance is not just a technical consideration—it's a crucial business strategy.

In this chapter, we will first examine the direct relationship between web performance and business conversions. You'll learn how site speed and user experience directly influence customer satisfaction, engagement, and ultimately, the conversion rates that drive business success. This understanding is vital for recognizing the tangible impact of performance on your bottom line.

Next, we'll explore the delicate balance between design complexity and site speed. In today's competitive digital landscape, a website needs to be visually compelling yet fast-loading. We'll navigate through strategies to achieve this balance, ensuring your site is both aesthetically pleasing and performance-optimized.

Understanding the indirect repercussions of subpar website performance is also crucial. We'll delve into how slow loading times, unresponsive pages, and downtime can negatively impact your brand reputation, search engine rankings, and customer loyalty, leading to long-term implications for your business.

Lastly, we will equip you with best practices for optimal web performance. From technical optimizations such as caching and image compression to strategic choices such as **content delivery networks (CDNs)** and responsive design, you'll learn a variety of tactics to enhance your website's speed and efficiency.

By the end of this chapter, you'll not only understand the importance of web performance in the broader context of your business goals but also possess the practical skills to implement effective optimization strategies. This knowledge is essential to anyone looking to solidify their web presence and drive business growth in the fast-paced digital world. Let's embark on this journey to master the art and science of web performance optimization.

The topics covered in this chapter are as follows:

- The direct link — performance and conversions
- Balancing design ambitions with load times
- The hidden costs of poor performance
- Strategies for a seamless web performance

The direct link — performance and conversions

In the digital landscape, the link between website performance and conversion rates is undeniable and significant. Understanding this relationship is crucial for businesses seeking to optimize their online presence for maximum return on investment. Performance, in this context, refers to how quickly and smoothly a website loads and functions, and conversions are the actions users take, such as making a purchase, signing up for a newsletter, or filling out a contact form.

Modern internet users have high expectations for website performance, including fast load times and seamless interactions. It has been shown that even a one-second delay in page response can lead to a significant drop in conversions. Users' attention spans are short, and if a website fails to load quickly, they are likely to leave and seek alternatives.

Website speed directly impacts the user experience. A fast-loading website provides a smooth, enjoyable experience that encourages users to explore more and engage deeper with the content. This positive experience increases the likelihood of users taking the desired conversion actions.

With the increasing use of mobile devices for internet access, mobile performance has become critical in the performance-conversions equation. Mobile users expect quick, responsive sites, and if a site isn't optimized for mobile performance, it can lead to a decrease in conversions from this significant user segment.

Search engines such as Google consider site speed and mobile-friendliness as ranking factors. Websites that perform better in these areas are more likely to rank higher in search results, leading to increased visibility, more traffic, and potentially higher conversion rates.

Fast and efficient websites are often perceived as more professional and trustworthy. Slow-loading sites or those that have performance issues can raise doubts about a company's credibility and quality, negatively impacting conversion rates.

To understand and improve the performance-conversions relationship, businesses need to analyze key website performance metrics such as load time, time to first byte, and interaction readiness. Tools such as Google Analytics and Google PageSpeed Insights can provide valuable insights into how performance impacts user behavior and conversions.

In summary, the performance of a website is intrinsically linked to its ability to convert visitors into customers or leads. By prioritizing and optimizing website performance, businesses can enhance user experience, improve search engine rankings, build trust, and ultimately drive higher conversion rates.

Next, we'll observe how the speed of a website is a key factor in user experience.

Speed as a key factor in user experience

In the digital age, where instant gratification is often the norm, the speed of a website or application plays a pivotal role in defining the overall user experience. Speed, in this context, is not just a measure of how quickly a page loads, but also how swiftly it responds to user interactions. This aspect of performance is critical, as it directly impacts how users perceive and interact with a digital platform.

The initial load time of a website sets the tone for the user's entire experience. A slow-loading site can frustrate users right from the start, increasing the likelihood of them leaving the site before even interacting with the content. Fast load times, on the other hand, create a positive first impression, encouraging further exploration.

Speed affects more than just the initial engagement. The responsiveness of a website or application as users navigate through it, click on links, fill out forms, or interact with various elements is crucial. Quick and responsive interactions foster a smooth and enjoyable experience, which is key to keeping users engaged and interested.

Speed is directly correlated with conversion rates. Users are more likely to complete a purchase, sign up for a service, or engage with content if they are not hindered by performance issues. In e-commerce, for instance, faster websites have been shown to have higher conversion rates and lower cart abandonment rates.

Beyond attracting new users, speed is also essential for retaining existing users. Consistently fast and reliable performance encourages users to return, building loyalty and trust over time. Slow speeds and lagging interactions, conversely, can drive even the most loyal customers away.

With the increasing use of mobile devices, the importance of speed is amplified. Mobile users often rely on cellular networks, which can be slower or less reliable than wired connections. Optimizing for mobile speed, therefore, becomes essential in providing a satisfactory user experience to a large and growing segment of internet users.

In conclusion, the speed of a website or application is a crucial determinant of the quality of user experience. It affects everything from first impressions to user engagement, conversion rates, and user retention. In a digital environment where users have a plethora of choices, ensuring optimal speed can be the difference between a successful online presence and one that is overlooked.

Now, let's discuss how performance can influence conversions.

Analyzing performance metrics that impact conversions

Understanding the direct link between website performance and conversion rates involves delving into specific performance metrics. These metrics provide tangible data points that can be analyzed to gauge how well a site performs and identify areas for improvement.

Here's a look at the key performance metrics that significantly impact conversions:

- **Page load time:** This is the time it takes for a page to fully display on a user's screen. It's a critical metric, as even a delay of a few seconds can lead to increased bounce rates and lost conversions. Tools such as Google PageSpeed Insights can measure page load time and offer recommendations for improvement.

- **Time to first byte (TTFB):** TTFB measures how long it takes for a user's browser to receive the first byte of data from the server. It's an indicator of server speed and a crucial component of overall page load time. A high TTFB can indicate server or network issues that need addressing.

- **First Contentful Paint (FCP):** FCP measures the time from navigation to when a user sees the first piece of content on the page. It gives an idea of how quickly a page becomes visually interactive. A faster FCP improves user perception of site speed.

- **First Input Delay (FID):** FID measures the time from when a user first interacts with your site (i.e., clicking a link, tapping on a button) to when the browser responds to that interaction. A low FID is crucial for a good user experience, especially for interactive sites such as online stores.

- **Cumulative Layout Shift (CLS):** CLS is a measure of visual stability. It quantifies how much visible content shifts around on the screen unexpectedly. A high CLS can be frustrating for users and lead to accidental clicks, potentially affecting conversions negatively.

- **Mobile performance:** Considering the increasing use of mobile devices, metrics related to mobile performance are also crucial. This includes mobile-specific load times, mobile responsiveness, and the functionality of interactive elements on touch screens.

- **Conversion path analysis:** Beyond general performance metrics, analyzing the user journey through the conversion path is vital. Understanding where in the process users drop off can pinpoint performance issues that are specific to certain pages or steps in the conversion funnel.

Regularly monitoring these performance metrics is essential in a data-driven approach to optimizing website performance. By correlating these metrics with conversion rates, businesses can make informed decisions and implement targeted optimizations to enhance both site performance and conversion potential.

Now that we understand the impact that performance has, let's talk about the strategies we can use to optimize performance.

Optimization strategies for enhanced performance

Optimizing website performance is a key strategy for enhancing user experience and, consequently, improving conversion rates. A website that loads quickly and operates smoothly retains users for longer and encourages them to engage more deeply.

Here are some essential strategies for boosting website performance:

- **Image optimization**: Large images can significantly slow down page load times. Optimizing images by compressing them and using appropriate formats (such as JPEG for photographs and PNG or SVG for graphics with transparent backgrounds) can drastically reduce load times. Consider implementing responsive images with the `srcset` attribute to serve different-sized images based on the user's device.

- **Minimizing HTTP requests**: Each element on a page (such as images, scripts, and CSS files) requires an HTTP request to load. Reducing the number of elements can speed up loading times. Techniques include using CSS sprites to combine images, merging multiple CSS and JavaScript files into single files, and eliminating unnecessary scripts and plugins.

- **Utilizing browser caching**: Caching stores elements of your website on the user's browser so that they don't need to be reloaded on subsequent visits. Setting up proper caching policies can significantly improve load times for returning visitors.

- **Enabling compression**: Use file compression tools for HTML, CSS, and JavaScript files. gzip compression is widely supported and can dramatically reduce the size of files being sent to the browser, enhancing load speed.

- **Using a CDN**: CDNs distribute your content across multiple servers around the world, so it's loaded from the server closest to the user. This reduces latency and speeds up access to your website.

- **Optimizing CSS and JavaScript**: The way CSS and JavaScript are loaded can impact performance. Use asynchronous loading for JavaScript to prevent it from blocking other elements from loading. Prioritize critical CSS to load first and defer less important styles to be loaded after the page renders.

- **Optimizing for mobile**: Ensure your website is mobile-friendly, with responsive design and mobile-optimized elements. This is particularly important given Google's mobile-first indexing approach.

- **Server performance optimization**: Choose a reliable hosting provider and consider upgrading to a better hosting plan if needed. Regularly monitor server health and performance. For dynamic websites, consider using a tool such as Varnish Cache to speed up content delivery.

- **Regular audits and monitoring**: Use performance monitoring tools such as Google PageSpeed Insights, Lighthouse, or GTmetrix to regularly check your website's performance and identify areas for improvement.

By implementing these optimization strategies, websites can significantly improve their performance, leading to a better user experience and higher conversion rates. It's important to remember that website optimization is an ongoing process and should be a key part of your web maintenance routine.

Now, let's explore some case studies.

Case studies — performance improvements and ROI

Real-world case studies often illustrate the direct impact of website performance improvements on **return on investment (ROI)** most effectively. By examining specific instances where enhancements in website speed and usability led to tangible business results, we can gain valuable insights into the relationship between performance optimizations and conversions.

Now, let's look into some case studies:

- **Case study 1: E-commerce retailer sees sales spike with faster load times**

    An online retailer specializing in outdoor gear implemented a series of performance optimizations, including image compression, browser caching, and server upgrades. The result was a 40% reduction in page load times. This improvement led to a 15% increase in page views, a 20% decrease in bounce rate, and, most significantly, a 10% increase in sales. The direct correlation between faster load times and increased sales demonstrated the substantial ROI of performance enhancements.

- **Case study 2: Travel website boosts bookings with mobile optimization**

    A travel booking company recognized a growing trend in mobile usage among its customers but had a site that was not fully optimized for mobile devices. After revamping its site with a mobile-first design, optimizing images, and implementing responsive layouts, it experienced a 50% increase in mobile traffic. More importantly, mobile bookings increased by 35%, directly impacting the company's bottom line and proving the value of mobile optimization.

- **Case study 3: Media site reduces server load, increases ad revenue**

    A popular news and media website struggled with slow loading times due to heavy content and high server load. By optimizing its content delivery with a CDN, implementing lazy loading for images and ads, and optimizing its backend infrastructure, the site improved its load time by over 60%. This led to a significant increase in user engagement, with a 25% increase in page views per session and a 15% increase in ad revenue, showcasing how performance optimization can positively impact both user experience and revenue.

- **Case study 4: SaaS company enhances user retention with performance tweaks**

    A **software as a service (SaaS)** company offering cloud-based tools noticed a high user drop-off rate during its onboarding process. Analysis revealed performance issues as a major contributor. By streamlining its application's loading process and optimizing critical scripts, the company reduced the average loading time by 30%. This change resulted in a 20% decrease in user drop-off and a substantial increase in long-term subscriptions, highlighting the importance of performance in user retention.

These case studies underscore the tangible benefits of investing in website performance optimization. Improvements in speed and user experience can lead to increased traffic, higher user engagement, better conversion rates, and ultimately, a significant ROI. They serve as compelling evidence for businesses to prioritize performance as a key element in their digital strategy.

In the next section, we'll explore how to best balance design ambitions with load times.

Balancing design ambitions with load times

In the digital world, the interplay between design ambitions and website performance creates a complex dilemma. On one hand, visually stunning designs with rich media and interactive elements can significantly enhance user experience and engagement. On the other hand, these design elements often come with the cost of increased load times and potentially decreased site performance. Navigating this dilemma is crucial for creating websites that are both aesthetically pleasing and functionally efficient.

Websites with ambitious designs, including high-resolution images, videos, intricate animations, and sophisticated graphical elements, offer a visually compelling experience. They can effectively convey a brand's identity and values, engage users emotionally, and differentiate the brand's website from competitors. However, these design elements are often data-heavy and can slow down the site.

Website performance, particularly load time, is a fundamental aspect of user experience. Slow-loading websites lead to higher bounce rates and can frustrate users, negating the positive impact of a visually appealing design. Users often value quick and seamless access to content over elaborate design features, especially when browsing on mobile devices.

The key challenge is to find a balance between implementing creative design elements and maintaining optimal site performance. This balance is crucial for ensuring both user satisfaction and engagement. It involves making strategic choices about which design elements are truly necessary and finding ways to optimize them for performance.

Various techniques can help mitigate performance issues while maintaining design integrity. This includes optimizing image and video sizes, using responsive and adaptive design practices, lazy loading media elements, and leveraging modern web technologies such as CSS3 and HTML5 that offer lightweight alternatives to traditional heavy graphics.

Regularly testing website performance and analyzing how different design elements impact load times and user behavior is vital for building a website that will last. Tools such as Google PageSpeed Insights and A/B testing can provide insights into how design changes affect performance and user engagement.

Ultimately, decisions around design and performance should be guided by user expectations and needs. This might involve tailoring the approach based on the target audience, as some user segments may prioritize rich visual design over speed while others prefer fast, streamlined experiences.

Balancing design ambitions with load times is an ongoing process, requiring continuous monitoring, optimization, and adjustment. By effectively navigating this dilemma, businesses can create websites that are both visually stunning and high-performing, catering to the needs and expectations of their users.

Next, we'll explore what the principles of efficient design are.

Principles of efficient design

Balancing the aesthetics of design with website performance necessitates adherence to certain principles that ensure efficiency. These principles guide the creation of visually appealing sites that also load quickly and perform well, thereby offering a superior user experience.

Here's a look at some of these essential principles:

- **Opt for simplicity**: A principle of efficient design is to embrace simplicity. This doesn't mean compromising on creativity or visual appeal but rather adopting a clean, uncluttered design. Simple designs with a clear message can be just as impactful as more complex ones, and they typically load faster and provide a better user experience.

- **Prioritize content**: Determine what content is most important to your users and make it the focus of your design. By streamlining content, you not only improve load times but also make your site easier to navigate, enhancing the overall user experience.

- **Optimize images and media**: Efficient design involves optimizing visual elements. Use compression tools to reduce file sizes of images and videos without sacrificing quality. Also, consider the appropriateness of the format — for example, using JPEGs for photographs and SVGs for icons and logos.

- **Use responsive design**: Responsive design is essential in creating efficient web designs. It ensures that your site looks good and functions well on any device, from desktops to smartphones. Responsive design adapts to the user's screen size and orientation, improving both performance and user experience.

- **Implement progressive loading**: Progressive loading, or lazy loading, is a technique where non-essential resources are loaded only when they are needed. For instance, images can be loaded as the user scrolls down the page rather than all at once. This improves initial page load times and reduces the amount of data used.

- **Leverage CSS and HTML5**: Modern CSS and HTML5 provide powerful tools for creating visually appealing designs without relying heavily on images and other heavy media files. These technologies allow for sophisticated styling and animations that are much lighter in terms of performance load.

- **Regular performance testing**: Continuously test the performance of your design. Use tools such as Google PageSpeed Insights to measure how design choices impact site speed and make adjustments accordingly. This ensures that your design remains efficient as technologies and user expectations evolve.

By adhering to these principles of efficient design, web developers and designers can create sites that strike a perfect balance between form and function. These principles help in crafting websites that not only captivate users with their aesthetic appeal but also provide them with a fast and smooth browsing experience.

Now that we understand the principles, let's explore techniques for optimization.

Techniques for optimizing visual elements

In the endeavor to balance design ambitions with load times, optimizing visual elements is key. This involves using various techniques to ensure that images, videos, and other design components contribute to an aesthetically pleasing design while not impeding the website's performance.

Here are some effective techniques for optimizing these visual elements:

- **Image compression and resizing**: Large image files can significantly slow down a website. Use image compression tools to reduce file sizes without noticeable loss in quality. Also, resizing images to the maximum display size needed on the site prevents unnecessarily large images from being loaded.

- **Choosing the right image format**: Different image formats have different strengths. For example, JPEG is great for photographs due to its high compression levels, while PNG is better for images requiring transparency. WebP, a newer format, offers high-quality images at smaller file sizes compared to both JPEG and PNG.

- **Implementing responsive images**: Use HTML's `srcset` attribute to serve different image sizes based on the user's screen size and resolution. This ensures that smaller images are sent to mobile devices, reducing load times and data usage.

- **Utilizing CSS sprites**: CSS sprites combine multiple images into a single image file, reducing the number of HTTP requests. This is particularly useful for icons and buttons that are used across the site.

- **Lazy loading of media**: Implement lazy loading for images and videos, which delays the loading of these elements until they are needed (usually when they are about to enter the viewport). This can significantly improve page load times, especially for pages with a lot of visual content.

- **Optimizing video content**: For video content, consider using compressed video formats and providing different resolutions. Host videos on external platforms such as YouTube and Vimeo to leverage their optimized playback technologies, which can be embedded into the site.

- **Efficient use of animations and interactions**: While animations and interactive elements can enhance the user experience, they can also impact performance. Use them judiciously and test their impact on load times. Consider using CSS3 animations, which are generally more performance-efficient than JavaScript-based animations.

- **SVGs for scalable graphics**: Use **Scalable Vector Graphics** (**SVGs**) for icons, logos, and other graphics. SVGs are resolution-independent and typically have smaller file sizes compared to raster images, making them ideal for responsive design.

- **Prioritizing above-the-fold content**: Optimize and prioritize the loading of visual content that appears above the fold (the portion of the web page visible without scrolling). This ensures that the user sees a fully loaded page quickly, improving the perceived performance.

By implementing these techniques, designers, and developers can ensure that their websites are visually rich without compromising on speed and performance. Regular testing and optimization should accompany these strategies to adapt to changing web standards and user expectations.

Up next, we'll explore how advanced web technologies have changed the world of web design.

Impact of advanced web technologies on performance

The integration of advanced web technologies has been a game-changer in web design and development, offering new possibilities for creating rich, interactive, and visually appealing websites. However, these technologies can have varying impacts on website performance. Understanding this impact is crucial for maintaining a balance between implementing cutting-edge design and ensuring optimal website speed and efficiency.

Here's a list of some of these advanced web technologies:

- **Advanced JavaScript frameworks and libraries**: Frameworks such as Angular, React, and Vue.js enable the creation of dynamic **single-page applications** (**SPAs**) with rich user interfaces and features. While these frameworks enhance user experience and interactivity, they can also add significant load to the website if not used judiciously. To mitigate performance issues, optimize the use of these frameworks by code splitting, lazy loading components, and minimizing the use of large third-party libraries.

- **CSS3 and HTML5 for design and animation**: CSS3 and HTML5 have brought a plethora of design capabilities such as transitions, animations, and complex layouts without heavy reliance on images or additional plugins. While they generally offer better performance than their JavaScript-based counterparts, overuse of complex CSS3 features or large HTML5 canvases can still impact the site's loading time and render speed.

- **Web fonts and typography**: Custom web fonts enhance visual appeal and brand identity but can negatively impact performance. Each font file adds to the overall page weight and requires additional HTTP requests. To optimize performance, limit the number of font variants used, leverage font-display swap in CSS, and consider using system or locally hosted fonts where appropriate.

- **High-definition media and retina displays**: Catering to high-resolution retina displays involves using higher-resolution images and graphics, which can increase page size. Employ techniques such as adaptive image serving, compression, and using vector graphics where possible to provide high-quality visuals without compromising load times.

- **Multimedia and interactive elements**: Rich multimedia content such as videos, interactive 3D models, and complex animations can significantly enhance user engagement but are often data-heavy. Optimizing these elements is key — compress video and audio files, use efficient encoding formats, and consider hosting videos externally to leverage advanced streaming technologies.

- **Web APIs and third-party integrations**: Integrating third-party services and APIs adds functionality but can also impact performance, especially if these services have slow response times or lead to additional requests. Ensure that third-party scripts are loaded asynchronously or deferred, and regularly monitor their impact on performance.

- **Progressive web apps: Progressive web apps** or **PWAs** use modern web capabilities to deliver app-like experiences. While they offer advantages such as offline support and faster subsequent loads, initial setup and caching strategies need to be carefully managed to prevent performance issues.

- **Testing and optimization**: Continuously test the performance impact of these technologies using tools such as Google Lighthouse or PageSpeed Insights. Regularly review and update the usage of these technologies to align with evolving web standards and performance best practices.

In summary, while advanced web technologies offer incredible opportunities for innovative web design, their impact on performance should be carefully managed. Striking the right balance involves making informed choices about technology usage, continuously monitoring their impact, and implementing optimization strategies to ensure a fast, efficient, and engaging user experience.

Now, let's get into some case studies.

Case studies — successful integration of design and speed

Exploring real-world case studies where businesses have successfully integrated ambitious design with high-speed performance can provide valuable insights into how this balance is achievable. These examples illustrate the impact of strategic design and optimization choices on both user experience and overall website effectiveness. Let's take a look:

- **Case study 1: Online fashion retailer revamps website design**

  An online fashion retailer faced challenges with slow load times due to high-resolution images and a complex layout. To address this, it implemented a series of optimizations:

  - Compressed and resized images without compromising quality

  - Implemented lazy loading for off-screen images

  - Switched to a more efficient, lightweight JavaScript framework

  - Adopted a mobile-first design approach

Post-optimization, the retailer saw a 30% decrease in bounce rates and a 20% increase in time spent on the site, leading to a 15% increase in conversion rates.

- **Case study 2: Tech company enhances user experience on product pages**

A tech company specializing in consumer electronics revamped its product pages, which initially suffered from slow loading due to interactive 3D models and high-definition videos. It made the following optimizations:

  - Optimized 3D models and deferred their loading until user interaction

  - Used compressed video formats and implemented adaptive streaming

  - Applied CSS3 for lightweight interactive elements instead of JavaScript-heavy widgets

  The result was a 50% improvement in page load times and a 25% increase in user engagement with the product pages, leading to higher sales.

- **Case study 3: News portal optimizes for speed and visual appeal**

A popular news portal struggled to balance its visually rich design with performance. The site was redesigned with the following optimizations:

  - Utilized CSS sprites to reduce HTTP requests for numerous icons and graphics

  - Implemented server-side rendering for faster content delivery

  - Prioritized above-the-fold content loading

  These changes led to a 40% improvement in load times and a significant reduction in server load, resulting in a better user experience and increased ad revenue.

- **Case study 4: Travel blog balances high-quality images with performance**

A travel blog with a global audience had a visually stunning design relying heavily on high-quality images, which impacted load times. Here is how they improved their performance:

  - Employed progressive image rendering and responsive image techniques

  - Optimized image file sizes using next-gen formats such as WebP

  - Redesigned the site's layout to be more efficient and lightweight

These optimizations led to a 60% improvement in loading speeds, a decrease in bounce rate by 35%, and a noticeable increase in page views and user session duration.

These case studies demonstrate that with thoughtful design choices and strategic optimization, it is possible to create websites that are both aesthetically pleasing and performant. The key lies in understanding the unique requirements of a site, regular performance testing, and staying updated with the latest web optimization practices.

In the next section, we'll dive into what the hidden costs of poor performance are.

The hidden costs of poor performance

In the digital age, the performance of a website or application is inextricably linked to **user experience** (**UX**), which in turn significantly impacts brand perception. A website's performance is often the first interaction a customer has with a brand, and poor performance can lead to negative perceptions that are hard to reverse. Understanding this relationship is key to prioritizing performance improvements.

Here are some factors to think about when analyzing performance:

- **First impressions matter**: The initial interaction a user has with a website sets the tone for their overall perception of the brand. Slow load times, unresponsive pages, and glitches create frustration and disappointment, casting a negative light on the brand. Conversely, a smooth, fast, and efficient online experience reflects positively on the brand's professionalism and reliability.

- **Performance as a reflection of quality**: Users often equate the performance of a website or application with the quality of the products or services offered. A high-performing site is seen as an indicator of a brand that values quality and attention to detail, whereas a poorly performing site can raise doubts about the brand's commitment to quality.

- **User experience and emotional connection**: UX goes beyond usability; it encompasses how users feel when interacting with a website. A seamless and enjoyable experience can foster positive emotions, strengthening the emotional connection between the user and the brand. This emotional resonance is crucial in building brand loyalty.

- **The ripple effect of negative experiences**: Negative experiences on a website can have a ripple effect. Dissatisfied users are likely to share their experiences with others, both online and offline, potentially damaging the brand's reputation. In the era of social media, a single negative experience can be amplified, reaching a wide audience rapidly.

- **Lost opportunities for engagement and conversion**: Poor performance hinders a website's ability to engage users effectively. Slow-loading content, broken links, or unresponsive design elements can cause users to leave the site prematurely, leading to lost opportunities for engagement and conversion.

- **Impact on customer retention**: Consistently poor website performance can lead to a decline in customer retention. Users who encounter repeated performance issues are more likely to seek alternatives, reducing the likelihood of repeat business and long-term customer relationships.

In summary, the performance of a website or application plays a critical role in shaping UX and, by extension, brand perception. Investing in performance optimization not only enhances user satisfaction but also contributes to a positive brand image, fostering trust, loyalty, and long-term customer relationships.

Now, let's break down how search engines play a role.

Search engine rankings and visibility

The performance of a website doesn't just affect UX; it also plays a crucial role in its visibility on search engines. Search engines such as Google increasingly prioritize site speed and UX factors in their ranking algorithms. Understanding the impact of website performance on search engine rankings and visibility is essential for any digital strategy.

Search engines aim to provide users with the best possible results, which include not only the relevance of content but also the quality of the UX. Google has explicitly stated that site speed is a ranking factor, meaning faster websites have a better chance of ranking higher in search results.

With the shift toward mobile-first indexing, Google predominantly uses the mobile version of a site for indexing and ranking. This change underscores the importance of mobile site performance. Websites not optimized for mobile can suffer in rankings, reducing their visibility.

Search engines also consider user engagement metrics such as bounce rates, time on site, and click-through rates. Slow-loading websites often have higher bounce rates and lower user engagement, which can signal to search engines that the site may not provide a good UX, potentially affecting its rankings.

Poor site performance can also impact technical SEO aspects. Slow load times can hinder search engine crawlers from effectively indexing a site. Additionally, performance issues can lead to an increase in crawl errors, further impacting SEO.

Google's Core Web Vitals are a set of specific factors that Google considers important in a web page's overall UX. Metrics such as **Largest Contentful Paint** (**LCP**), FID, and CLS are now integral to Google's ranking criteria, directly tying site performance to SEO.

The impact of performance on SEO is not isolated to a single factor; it's a compound effect. Poor performance can lead to lower rankings, which in turn leads to reduced visibility and traffic, further impacting the site's ability to rank well.

To maintain and improve search engine rankings, continuous optimization and monitoring of website performance are necessary. Regularly auditing site speed, mobile usability, and Core Web Vitals, and addressing any issues is crucial for keeping up with search engine algorithms and user expectations.

In conclusion, the performance of a website is a key determinant of its success in search engine rankings and online visibility. As search engines evolve to prioritize UX, optimizing site performance becomes increasingly important for anyone looking to improve their SEO and online presence.

Now, let's dive into more complex topics such as bounce rates and user retention.

Bounce rates and user retention challenges

Poor website performance not only affects immediate UX but also has a profound impact on bounce rates and long-term user retention. A website's bounce rate reflects the percentage of visitors who leave the site after viewing only one page, and it is a critical metric for understanding user engagement and satisfaction. High bounce rates and challenges in user retention can often be attributed to underlying performance issues.

The first few seconds of a user's experience on a website are crucial. If a site takes too long to load, users are likely to become impatient and leave, leading to a high bounce rate. This initial impression is pivotal in either capturing the user's interest or losing it entirely.

Numerous studies have established a direct correlation between load times and bounce rates. Even a delay of a few seconds can significantly increase the likelihood of users abandoning the site. This is particularly true for mobile users who often rely on varying network speeds.

Slow or unresponsive websites disrupt the UX, making navigation, reading, and interaction cumbersome. This not only contributes to high bounce rates but also affects overall user engagement with the content and functionalities of the site.

Consistently poor performance leads to diminished user trust and satisfaction, which are crucial for long-term retention. Users who repeatedly face performance issues are less likely to return, affecting the website's ability to build a loyal user base.

High bounce rates directly impact conversion rates. If users leave a site prematurely due to performance issues, the chances of them converting — whether that's making a purchase, signing up for a newsletter, or completing a contact form — are significantly reduced.

High bounce rates can also have indirect effects on SEO. Search engines interpret high bounce rates as a signal that a website might not be offering relevant or satisfactory content to its visitors, which can negatively impact search rankings.

To combat high bounce rates, focus on optimizing load times through techniques such as image compression, minimizing HTTP requests, using CDNs, and optimizing code. Additionally, improving the overall UX by designing intuitive navigation and engaging content layouts can encourage users to explore the site beyond the initial landing page.

In summary, bounce rates and user retention are significantly influenced by website performance. By addressing performance issues, not only can immediate bounce rates be reduced, but it also sets the foundation for enhanced user satisfaction and loyalty, which are crucial for the long-term success of any online presence.

What would the impact of poor conversion rates look like?

Conversion rate reductions and lost revenue

The hidden costs of poor website performance significantly impact conversion rates and ultimately lead to lost revenue. In today's fast-paced digital environment, where user patience is limited and alternatives are just a click away, the efficiency of a website is a critical determinant of its commercial success. Understanding how performance issues translate into conversion rate reductions and financial losses is essential for businesses relying on online platforms.

There is a well-established connection between website speed and conversion rates. Delays in page loading, even by a few seconds, can frustrate users and lead them to abandon their shopping carts or exit the website altogether. Each abandoned transaction directly translates to lost revenue.

A seamless UX is key to guiding customers through the purchase process. Slow loading times, unresponsive pages, and checkout process delays not only hinder this journey but also erode trust in the brand. This can result in users reconsidering their purchase decisions, decreasing conversion rates.

Poor performance doesn't just affect immediate sales; it also impacts the long-term value of customers. Negative experiences can deter customers from returning, thereby reducing the potential lifetime revenue they could bring to the business.

The transition from browsing to purchasing is delicate and can be easily disrupted by performance issues. Websites that are quick and responsive encourage users to move smoothly from browsing products or services to making a purchase, thereby increasing the likelihood of conversion.

Reduced website visibility in search engine rankings due to poor performance can lead to decreased organic traffic. Less traffic means fewer opportunities to convert visitors into customers, directly impacting revenue.

In competitive markets, brand perception is key. Websites that perform poorly may be perceived as less credible or lower quality, adversely affecting market positioning. This can have long-term implications for brand reputation and revenue generation.

Tools such as Google PageSpeed Insights can help quantify the impact of performance on conversions. They provide insights into how improving load times could increase revenue. For instance, studies have shown that improving site speed by just one second can significantly boost conversion rates.

To mitigate these challenges, focus on optimizing website performance through techniques such as image and video optimization, efficient coding, server upgrades, and using a CDN. Additionally, streamline the user journey, especially the checkout process, to minimize any obstacles that could lead to cart abandonment.

In conclusion, the hidden costs of poor performance, manifesting as reduced conversion rates and lost revenue, are significant. By prioritizing website performance optimization and continuously monitoring and addressing issues, businesses can enhance user experiences, boost conversions, and secure their revenue streams.

Now, let's explore the long-term implications.

Long-term implications for business growth

The hidden costs of poor website performance extend far beyond immediate metrics such as bounce rates and conversion dips; they have profound long-term implications for overall business growth. In an increasingly digital-centric market, the performance of online platforms can significantly influence a business's ability to expand, attract new customers, and retain existing ones. Understanding these long-term implications is crucial for strategic planning and sustained business success.

A website that consistently underperforms not only deters potential new customers but also risks losing existing ones. Slow load times and poor user experiences can tarnish a brand's image, making both acquisition and retention more challenging and expensive.

Online experiences shape brand perception. Performance issues can lead to negative reviews and social media feedback, harming a brand's reputation. Rebuilding trust and credibility in the digital space can be a lengthy and complex process, impacting long-term business prospects.

In competitive markets, website performance can be a key differentiator. Companies with optimized, high-performing websites are more likely to outpace competitors with slower, less reliable sites. Poor performance can thus place businesses at a significant disadvantage, hampering growth and market share expansion.

As businesses grow, their websites typically experience increased traffic and higher demand. If performance issues are not addressed early on, scaling up can exacerbate these problems, leading to even greater losses in terms of user satisfaction and revenue.

The effectiveness of digital marketing campaigns is closely tied to website performance. Slow loading times can diminish the returns on paid advertising and SEO efforts as potential customers are less likely to engage with a poorly performing site.

Addressing performance issues often requires investing in better infrastructure, optimization tools, and expert personnel. Neglecting performance can lead to higher costs in the long run, as more resources are needed to fix deep-rooted problems and recover lost ground.

Poor website performance can also stifle innovation. Resources that could be used for developing new features, services, or content may be diverted to address performance issues, slowing down a business's ability to adapt and innovate in a rapidly changing digital landscape.

For businesses aiming to expand globally, website performance is critical, especially in regions with slower internet speeds. Performance optimization is key to ensuring accessibility and a positive UX worldwide, which is crucial for international growth.

In summary, the long-term implications of poor website performance are extensive and can affect nearly every aspect of a business. Proactively addressing these issues is not just about immediate gains but is also fundamental to securing sustainable business growth, maintaining a competitive edge, and fostering a strong, reputable brand in the digital era.

Let's dive into strategies that can be used to help with a seamless web performance.

Strategies for a seamless web performance

Web performance optimization (WPO) is crucial for ensuring a seamless UX and bolstering a website's effectiveness. Adhering to the core principles of WPO can significantly enhance site speed, reliability, and usability. Understanding and implementing these principles is essential for any business aiming to establish a strong online presence.

Here are several strategies that should be considered:

- **Prioritizing UX**: The primary goal of WPO is to improve UX. This means ensuring the site is responsive, intuitive, and fast. Every optimization decision should be made with the user in mind, from reducing load times to simplifying navigation.

- **Optimizing load times**: The speed at which pages load is critical. Techniques such as compressing images and videos, minifying CSS and JavaScript files, and leveraging browser caching are fundamental in reducing load times. Fast-loading pages reduce bounce rates and increase user engagement.

- **Implementing responsive design**: With the variety of devices used to access the web, responsive design is essential. A website must perform well across all devices — desktops, tablets, and smartphones. Responsive design ensures websites adapt their layouts and content to fit different screen sizes and resolutions.

- **Streamlining code and content**: Efficient coding is key. This involves cleaning up HTML, CSS, and JavaScript to remove unnecessary characters, comments, and unused code. Streamlining content also means organizing and structuring it in a way that enhances performance and UX.

- **Using CDNs**: CDNs distribute website content across multiple servers around the world. This reduces the distance between the server and the user, enhancing load times, especially for media-heavy content.

- **Optimizing server performance**: The web server plays a vital role in performance. Ensuring the server is well-configured and using the latest technologies can greatly improve response times. This includes regular updates, efficient database management, and using server-side caching.

- **Monitoring and analyzing performance**: Continuous monitoring and analysis of web performance are essential. Tools such as Google PageSpeed Insights, Lighthouse, and WebPageTest can provide valuable insights. Regularly testing and adapting to the evolving web landscape is crucial for maintaining optimal performance.

- **Incorporating progressive enhancement**: This strategy involves creating a basic functional experience for all users and then enhancing it for more capable browsers or devices. It ensures that the website remains accessible and usable, regardless of the user's device or network conditions.

- **Balancing between design and functionality**: While aesthetics are important, they should not compromise performance. Finding the balance between a visually appealing design and a fast, responsive site is crucial. This might involve optimizing images, using web fonts judiciously, and avoiding heavy design elements that slow down the site.

By adhering to these core principles of web performance optimization, businesses can create websites that are not only visually appealing but also perform efficiently. This leads to a better UX and improved search engine rankings, and ultimately contributes to the success of the online aspects of the business.

We've already spoken about images and media content, but let's see how we can optimize them to aid in performance.

Optimizing images and media content

Optimizing images and media content is a crucial aspect of web performance strategy. Since visual elements typically consume the most data on websites, their optimization can significantly improve loading times, enhance UX, and contribute positively to SEO.

Here are some essential strategies for effectively optimizing images and media:

- **Image compression**: Reducing file sizes without losing significant quality is key. Tools such as Photoshop, GIMP, and online services such as TinyPNG and ImageOptim can compress images efficiently. Compression works by removing unnecessary data, resulting in a lighter file that loads faster.

- **Choosing the right format**: Selecting the appropriate format for each type of image is crucial. JPEG is suitable for photographs due to its high compression. PNG is ideal for images requiring transparency. For a balance of quality and compression, WebP is increasingly popular and widely supported.

- **Implementing responsive images**: Use HTML's `srcset` and `sizes` attributes to serve different image sizes for various screen resolutions. This ensures that users download only the version of the image appropriate for their device, saving bandwidth and improving load times.

- **Lazy loading techniques**: Lazy loading delays the loading of images and media until they are needed (typically when they enter the viewport). This can significantly reduce initial page load times, especially for pages with a large number of images.

- **Image sprites for icons**: Combine multiple icons and small images into a single sprite sheet. This reduces the number of HTTP requests as the browser loads one image file instead of many small ones. CSS can be used to display only the relevant portion of the sprite for each icon.

- **Video optimization**: Compress video files and offer different resolutions to cater to various internet speeds and devices. Hosting videos on external platforms (such as YouTube and Vimeo) and embedding them can also improve performance as these platforms are optimized for fast streaming.

- **Optimizing animated content**: For animations, consider file size and playback performance. CSS animations are often more performance-friendly compared to GIFs or heavy JavaScript animations.

- **Using thumbnails for larger media**: Utilize thumbnails for large images and videos, allowing users to decide whether they want to load the full-size media. This can improve page load times and save bandwidth for users who may not view every piece of media.

- **Regular performance reviews**: Continuously monitor and review the performance impact of images and media. As website content gets updated and added, regular optimization ensures ongoing performance efficiency.

In conclusion, by applying these optimization techniques, websites can significantly enhance their loading speed and performance. This not only improves the UX but also positively impacts search engine rankings, making it a critical component of a successful web strategy.

Another consideration would be caching and using CDNs.

Leveraging caching and CDNs

To achieve seamless web performance, leveraging caching and CDNs is essential. These technologies play a pivotal role in reducing load times, handling high traffic loads, and delivering content efficiently to users worldwide.

Caching involves storing copies of files in a temporary storage location known as a cache. This allows for quicker access to frequently requested data and reduces the load on the server. There are various types of caching, including browser caching, server caching, and CDN caching. Let's look at a few:

- **Browser caching**: This stores website resources on the user's device. When the user revisits the website, the browser can load the page from the local cache instead of downloading everything again.

- **Server caching**: Implemented on the server side, it stores the rendered page so that it doesn't need to be rebuilt for each request. Technologies such as Varnish can be used for server caching.

CDNs are networks of servers strategically located around the globe, designed to deliver web content efficiently to users from the server nearest to them. By caching content on these servers, CDNs reduce latency, speed up content delivery, and handle high traffic loads more effectively. Here are some benefits of CDNs:

- **Global reach and speed**: CDNs are especially beneficial for websites with an international audience. They ensure that content is delivered quickly, regardless of the user's geographic location.

- **Reduced server load**: By serving content from CDN servers, the load on the original server is reduced, which can significantly improve site performance during peak traffic times.

To maximize the benefits of caching, implement strategies such as setting appropriate expiry times for different types of content. Static resources such as images and CSS files can have longer cache times, while dynamic content may require shorter cache times or no caching.

Many CDN services offer additional features such as optimization of images and videos in real-time, HTTPS support for secure delivery, and DDoS (distributed denial of service) protection. Popular CDN providers include Akamai, Cloudflare, and Amazon CloudFront.

Faster load times and reduced server downtime achieved through effective caching and CDNs, can contribute to better search engine rankings. Google and other search engines favor websites that provide a faster and more reliable UX.

Regularly monitor the performance of caching and CDN implementations. Keep up with updates and changes in CDN technology to ensure your website continues to deliver content as efficiently as possible.

In conclusion, the strategic use of caching and CDNs is fundamental in optimizing web performance. These technologies not only enhance the UX by reducing load times and improving reliability but also contribute to better handling of traffic surges and improved SEO, key factors in the overall success of any online presence.

Now, let's talk about minimization.

Minimizing and compressing resources

A critical strategy in achieving seamless web performance is minimizing and compressing resources. This process involves reducing the size of website files, such as HTML, CSS, JavaScript, and media files, to enhance loading speeds and improve overall site efficiency. Efficient handling of these resources can lead to significant improvements in website performance.

Minification is the process of removing unnecessary characters (such as white space, comments, and newline characters) from code without changing its functionality. Tools are available for minifying HTML, CSS, and JavaScript files. This reduces file size, leading to faster transmission over the network and quicker parsing by browsers.

Beyond minification, CSS and JavaScript files can be further optimized. Techniques include removing unused CSS rules and JavaScript functions, using shorthand coding methods, and combining multiple files into one to reduce HTTP requests. Tools such as PurifyCSS and UglifyJS can automate much of this process.

Images often constitute the bulk of a web page's size. Compressing images reduces their file size without noticeably affecting quality. Numerous tools and services are available for both lossy and lossless compression. Choosing the right compression level and format (such as JPEG, PNG, or WebP) based on the image content and usage is crucial.

Similar to images, videos, and other multimedia elements should be compressed for web use. Using modern, efficient codecs and considering adaptive bitrate streaming for videos can significantly reduce the size of multimedia files.

Enabling gzip or Brotli compression on the server can greatly reduce the size of text-based resources as they are sent to the browser. This server-level compression can be enabled through web server configuration.

Web fonts can add significant weight to web pages. Optimize font loading by limiting the number of font variations used, using font-display swap to ensure text remains visible during font loading, and considering system fonts as alternatives where appropriate.

Third-party scripts, such as those used for analytics, social media integration, and advertising, can significantly impact performance. Audit these scripts regularly, remove non-essential ones, and defer the loading of others to reduce their impact on initial page load times.

Regularly audit the website's resources using tools such as Google PageSpeed Insights or WebPageTest to identify opportunities for minimization and compression. Keeping up with the latest best practices in resource optimization is key to maintaining optimal performance.

In summary, minimizing and compressing resources is a fundamental aspect of web performance optimization. This practice not only improves loading times but also enhances the UX, potentially leading to higher user engagement, retention, and conversions.

The next consideration would be adaptive loading.

Responsive design and adaptive loading

In the pursuit of seamless web performance, integrating responsive design with adaptive loading is paramount. This approach ensures not only that a website looks and functions optimally across different devices but also that it loads efficiently by adapting to the user's device and network conditions. This strategy is crucial for balancing aesthetics, functionality, and performance.

Responsive design refers to the practice of creating web pages that adapt to the size and orientation of the user's device. This involves using fluid grid layouts, flexible images, and CSS media queries to provide an optimal viewing experience across a range of devices, from desktops to smartphones.

While responsive design adjusts the layout, adaptive loading optimizes performance. It involves serving different assets or content based on the user's device capabilities and network conditions — for example, serving lower-resolution images to users on a slow mobile network or a less powerful device.

Media queries are a cornerstone of responsive design. They allow you to apply different CSS styles based on device characteristics, such as screen width, height, and orientation. This ensures that your website not only looks good but also performs well across all devices.

Adaptive loading can be implemented by conditionally loading resources. This includes deferring the loading of non-critical resources, such as images and videos, and loading them only when they are likely to be viewed by the user (often referred to as "lazy loading").

Beyond images and videos, adaptive loading can also apply to other content types — for instance, serving simpler, less resource-intensive animations or interactive elements to users on mobile devices, ensuring smooth performance even on less powerful hardware.

Responsive design for mobile devices also involves optimizing for touch interactions. This includes designing larger touch targets, simplifying interactions, and avoiding hover-based UI (user interface) elements that don't translate well on touch screens.

It's essential to test your website's performance across a range of devices and network conditions. This helps in identifying any potential issues with responsive design or adaptive loading, ensuring a consistent and smooth UX.

The challenge in integrating responsive design with adaptive loading lies in balancing visual aesthetics with performance. It requires careful consideration of which elements are essential for the UX and finding ways to optimize them for different conditions.

In conclusion, responsive design coupled with adaptive loading forms a powerful approach to web performance optimization. It not only makes websites accessible and user-friendly across various devices but also ensures they load efficiently, providing a seamless UX regardless of device or network conditions.

Summary

As we conclude this chapter, let's recap the essential skills and lessons we've covered, underscoring their significance in the realm of web development and business strategy.

We began by exploring the crucial relationship between web performance and business conversions. Understanding this connection is vital as it directly impacts user satisfaction, engagement, and ultimately, conversion rates. You've learned how every second of delay in page loading can affect customer behavior, potentially leading to lost sales and reduced revenue.

Balancing design complexity with site speed was another key focus of this chapter. We navigated the challenges of creating visually appealing websites without compromising on performance. This skill is critical in today's web landscape, where user expectations for both aesthetics and speed are high.

We also delved into the indirect repercussions of subpar website performance. You gained insights into how performance issues can negatively affect brand reputation, search engine rankings, and customer loyalty. This understanding highlights the broader implications of web performance beyond immediate user interactions.

Lastly, this chapter equipped you with best practices for optimal web performance. From technical optimizations such as image compression and leveraging browser caching to strategic implementations such as using CDNs and adopting responsive design, these practices are designed to enhance the overall efficiency and UX of your website.

These lessons are invaluable as they transcend basic web development, touching on aspects crucial to the growth and success of any online business. Implementing these strategies will not only improve your website's performance but also contribute significantly to achieving your broader business objectives.

In the next chapter, we will build on these concepts and focus on responsive design in a more granular fashion. You'll learn about strategically setting breakpoints in your design to ensure optimal adaptability across various devices and screen sizes. This is a natural progression from optimizing performance, as it delves deeper into how responsive design can be fine-tuned to meet diverse user needs and enhance the overall effectiveness of your digital presence. Prepare to deepen your understanding of responsive design and its critical role in modern web development.

6

Strategic Breakpoints and Adaptability

Welcome to *Chapter 6*, where we delve into the art of RWD in today's diverse digital environment. This chapter is designed to sharpen your understanding of current device usage trends and the methodology behind selecting effective breakpoints in web design.

You'll learn to distinguish the fine line between creating a design that's adaptively responsive and one that's unnecessarily complex. This balance is crucial in maintaining user engagement and ease of maintenance. We'll also explore the practical impacts of breakpoint decisions, highlighting how these choices affect user experience and website performance.

By the end of this chapter, you'll be equipped with the essential skills to craft websites that are visually appealing and functionally robust across various devices. This journey will enhance your ability to create adaptable, user-centered web experiences, which are essential in today's technology-driven world.

The topics covered in this chapter are as follows:

- Understanding the device and user landscape
- The art and science of choosing breakpoints
- Adaptability versus overcomplication
- Real-world implications of breakpoint choices

Understanding the device and user landscape

In today's rapidly evolving digital world, the landscape of devices and screens is continuously shifting, presenting both challenges and opportunities for web designers and developers. Understanding this evolving landscape is crucial for creating web experiences that are not only relevant but also accessible and enjoyable for all users, regardless of their choice of device.

The range of devices used to access the internet has expanded dramatically. No longer limited to desktop computers, users now routinely switch between smartphones, tablets, laptops, and even smart TVs to browse the web. Each of these devices has different screen sizes, resolutions, and capabilities, necessitating a more adaptable approach to web design.

The most significant shift in recent years has been the increasing dominance of mobile devices. Statistics show that mobile internet usage has surpassed desktop, a trend that is only growing. This shift has major implications for web design, underscoring the need for mobile-first approaches that prioritize the mobile user experience.

Beyond traditional devices, new types of screens are emerging, such as foldable phones and large-format displays. These innovative devices introduce unique design considerations, such as variable aspect ratios and the need for responsive layouts that can adapt to different screen states (such as folded and unfolded).

responsive web design (**RWD**) has evolved from a recommended practice to an absolute necessity. It's no longer about designing for a few standard screen sizes but about creating flexible layouts that can adapt to any screen. This shift requires an understanding of fluid grids, flexible images, and media queries.

It's also important to consider how user behavior varies across different devices. Desktop users might seek a more comprehensive browsing experience with multiple tabs and extensive navigation options, while mobile users often expect quick, accessible information with easy-to-tap buttons and streamlined content.

Finally, anticipating future trends in device usage is key. With technologies such as **augmented reality** (**AR**) and **virtual reality** (**VR**) gaining traction, forward-thinking designers are considering how these technologies will further diversify the device landscape and what that means for user interaction and web design.

In conclusion, the evolving landscape of devices and screens requires web designers and developers to be versatile, innovative, and user-centric in their approach. Understanding this dynamic landscape is the first step in creating web experiences that are not just functional but also delightful, regardless of how or where they are accessed.

User behavior across different devices

Understanding user behavior across different devices is fundamental for creating a web design strategy that effectively responds to the varied ways people interact with digital content. Different devices cater to different contexts, uses, and user expectations, which significantly influence how web content should be structured and presented.

Here is a list of various ways users interact with devices:

- **Context-specific usage**: The way users interact with devices often depends on their context. Desktops, often used in work or study environments, lend themselves to more in-depth browsing and multitasking. Mobile devices, on the other hand, are typically used on the go, calling for quicker access to information and simpler, more streamlined interactions. Tablets often bridge these two, being used both for leisure and productivity.

- **Differing interaction models**: Interaction models differ substantially across devices. Desktop users have the precision of a mouse and the versatility of a keyboard, allowing fine control and complex interactions. Mobile users rely on touch, necessitating larger, more accessible touch targets and gestures such as swiping. These differences should shape the navigation design and the overall interface layout.

- **Content consumption patterns**: Content consumption varies significantly between devices. Mobile users often seek quick answers or brief entertainment, favoring bite-sized content, easily digestible lists, and straightforward navigation. Desktop users might be more inclined to engage with longer-form content, detailed explorations, and complex interactive experiences.

- **Adapting to attention spans and speed**: Attention spans and browsing speeds differ across devices. Mobile users typically have short attention spans due to the on-the-go nature of mobile browsing. Websites accessed on mobile devices need to load quickly and present key information upfront. Desktop websites can afford to be more expansive, both in content depth and layout.

- **Task-oriented versus exploratory use**: Mobile users are often task-oriented, seeking specific pieces of information or completing particular actions such as shopping or booking. In contrast, desktop users might be more exploratory, browsing through content at a more leisurely pace. This distinction can inform how content and calls to action are prioritized and presented.

- **Responsive strategies**: In responsive design, understanding these behavioral differences is crucial for deciding how content reflows and resizes across devices. It's not just about making content fit on different screens but also about tailoring the user experience to match the typical behaviors and needs of users on those devices.

In conclusion, recognizing and adapting to user behavior across different devices is key to effective web design. By considering how users interact with content in various contexts, designers can create more intuitive, engaging, and successful web experiences. This user-centric approach is essential for meeting the diverse needs of today's digital audience.

Analyzing data for device usage insights

In the realm of strategic breakpoints and adaptability, analyzing data to understand device usage patterns is crucial. This insight informs how web designs can be optimized for various devices, ensuring that users receive the best possible experience. Data analysis helps in making informed decisions rather than relying on assumptions or general trends.

Here are the different ways to analyze data:

- **Understanding the audience**: The first step is to analyze your website's audience and their device preferences. Web analytics tools such as Google Analytics provide comprehensive data on user devices, including the types of devices accessing your site, their screen sizes, operating systems, and browsers. This data gives a clear picture of the most commonly used devices by your audience.

- **Device segmentation**: Segmenting this data can reveal important trends. For instance, you might find that the majority of your traffic comes from mobile devices, or that a significant portion of your audience uses tablets. Understanding these segments allows targeted design strategies that cater specifically to the majority of your users' needs.

- **Analyzing user behavior by device**: Beyond just identifying the devices, it's important to analyze how user behavior varies across these devices. Look into metrics such as session duration, bounce rate, and conversion rates for users on different devices. This can indicate whether your site currently meets the needs of users on these devices or if there are areas for improvement.

- **Impact of device usage on design decisions**: With this data, you can make more informed decisions about setting breakpoints in your responsive design. For example, if a significant portion of your audience uses mid-sized tablets, you might set a breakpoint that optimizes specifically for that screen size, ensuring those users have an optimal experience.

- **Adapting to changing trends**: Device usage trends can change over time, so it's important to continually monitor and analyze this data. Regularly updating your understanding of how your audience accesses your site will help you keep your web design current and effective.

- **Incorporating user feedback**: Combining quantitative data with qualitative feedback, such as user surveys or feedback forms, can provide a more holistic view of how well your website serves different device users. This can highlight specific issues that might not be evident from analytics alone.

In conclusion, analyzing data for device usage insights is a key skill in developing a strategic approach to breakpoints and web design adaptability. By understanding the devices your audience uses and how they interact with your site on these devices, you can create a more effective, user-friendly website that caters to the specific needs of your audience, ultimately enhancing user engagement and satisfaction.

Next, we'll explore how to best go about choosing breakpoints.

The art and science of choosing breakpoints

Understanding the fundamentals of breakpoint selection is essential for creating RWDs that fluidly adapt to various screen sizes. Breakpoints are the points at which a website's content and layout will adjust to provide the best user experience. Selecting the right breakpoints is both an art and a science, requiring a blend of technical knowledge and creative intuition. Here's a framework for creating breakpoints:

- **Defining breakpoints**: A breakpoint is a media query in CSS where certain styles are applied to make the content look appropriate on different devices. Essentially, they are the screen widths (measured in pixels) at which the website's design and layout will change to accommodate different screen sizes.

- **Purpose of breakpoints**: The primary purpose of breakpoints is to ensure that a website is usable and aesthetically pleasing across a range of devices. This involves adjusting elements such as navigation menus, images, grid layouts, and font sizes so that they are appropriate for the screen size and resolution.

- **Common breakpoint standards**: There are some standard breakpoints that are commonly used in responsive design, typically based on the screen sizes of popular devices:

  - Mobile devices (small screens): 480 px or less

  - Tablets (medium screens): 481 px to 768 px

  - Laptops/small desktops (large screens): 769 px to 1,024 px

  - Desktops (extra-large screens): 1,025 px and above

  These standard breakpoints serve as a starting point. However, they should be adapted based on the content and audience of the website.

- **Content-first approach**: An effective strategy for selecting breakpoints is the content-first approach. This involves looking at the content of your website and determining where it naturally breaks. Rather than forcing content to fit into device-based breakpoints, this approach ensures that the design adapts to the content, leading to a more natural and user-friendly experience.

- **Testing and adjustment**: Choosing breakpoints isn't a one-time task. It requires ongoing testing and adjustment. Use tools such as browser developer tools to test how your website scales on different screen sizes. Pay attention to how content flows, and adjust breakpoints as needed to ensure the layout remains intuitive and accessible.

- **Avoiding the overuse of breakpoints**: While it's important to cover a range of devices, overusing breakpoints can lead to unnecessary complications in your code. Aim for a balance where your website is responsive and adaptable without being bogged down by excessive media queries.

In summary, selecting the right breakpoints is a fundamental aspect of RWD. It requires an understanding of standard device dimensions and a deep consideration of how your specific content behaves across various screen sizes. By focusing on these fundamentals, you can create a website that provides an optimal viewing experience on any device.

Analyzing content and layout for breakpoint decisions

In RWD, determining where to set breakpoints is critical for ensuring that the content and layout adapt gracefully across different devices. The key to effective breakpoint decisions lies in a thorough analysis of your website's content and layout. This analysis identifies the points at which the design should adjust to maintain usability and aesthetic appeal.

Start by examining how your content flows on various screen sizes. Pay attention to elements such as text, images, and call-to-action buttons. Look for points where the content feels cramped, stretched, or loses its visual impact. These are indicators that a breakpoint might be needed to rearrange or resize content for a better user experience.

Evaluate your site's layout at different widths. Notice how multi-column layouts behave as the screen narrows. Do the columns stack effectively on smaller screens? Is the information hierarchy maintained? Sometimes, a layout that looks perfect on a desktop may become disorganized or lose its intuitive navigation on smaller screens. These observations are crucial in determining where breakpoints are necessary.

Identify critical viewport widths — the points where your design starts to break down or look less appealing. Resize your browser window and observe the changes in your website's layout. Make note of the widths where you need to modify your CSS to improve the presentation. These widths are your potential breakpoints.

On smaller screens, prioritizing key content is essential. During your analysis, decide which elements need to be highlighted and which can be condensed or moved. For example, secondary information might be hidden behind menus or accordions on mobile views to keep the focus on primary content.

Use real content instead of placeholder text and images to make more accurate decisions. Real content will give you a better sense of how much space is needed and how it flows on the page. This approach prevents surprises post-launch when the site is populated with actual content.

Consider how images and other media will resize and reflow. Large images may need to be scaled down or replaced with more suitable ones for smaller screens. Ensure that media enhances the content at each breakpoint, rather than distracting or overwhelming it.

Breakpoint decisions should be part of an iterative design process. Set initial breakpoints based on your analysis, then test and refine them. Use user feedback and analytics data to make further adjustments. This iterative approach ensures that your breakpoints effectively serve your diverse user base.

In conclusion, analyzing content and layout for breakpoint decisions is a meticulous process that requires attention to detail and an understanding of how different elements interact within a responsive design. By carefully examining your content flow, layout, and the behavior of media across screen sizes, you can set breakpoints that create a cohesive, engaging, and functional experience for all users.

Responsive design best practices and breakpoints

When implementing responsive design, understanding and applying best practices for breakpoints is essential. Breakpoints should not only respond to the width of the devices but also enhance the user experience. This section focuses on the best practices for setting breakpoints in responsive design, ensuring that your website adapts gracefully and effectively to various screen sizes.

Here's a list of some best practices:

- **Start with mobile-first design**: A mobile-first approach means designing for the smallest screen first and then scaling up to larger screens. This practice often leads to more optimized and performant designs as it forces a focus on essential content and functionality from the outset.

- **Use major device categories as guides**: While it's impossible to cater to every device individually, using major device categories (such as mobile, tablet, small desktop, and large desktop) as a starting point can be effective. These categories help in defining a broad range of breakpoints that cater to most users.

- **Focus on content, not just devices**: Breakpoints should be based on the content and design of your website rather than trying to match specific devices. Resize your browser to see where the content naturally breaks or looks awkward and set your breakpoints there. This approach ensures that your design is truly responsive to the user's needs.

- **Employ fluid grids**: Use fluid grid layouts that utilize percentages rather than fixed units. This allows your layout to flexibly adapt to different screen sizes, creating a more seamless transition between breakpoints.

- **Implement flexible images and media**: Ensure that images and other media elements are flexible. They should be able to resize within their containing elements to prevent overflow or distortion. Techniques such as CSS's **max-width: 100%** can be useful here.

- **Minimize the number of breakpoints**: While it's important to have a responsive site, too many breakpoints can overcomplicate your CSS and maintenance efforts. Strive for the minimum number of breakpoints that still provide an optimal user experience.

- **Prioritize accessibility**: Ensure that your responsive design maintains accessibility standards across all breakpoints. Elements such as navigation menus, buttons, and form inputs should be easily usable on all devices.

- **Test on real devices**: While browser resizing tools are helpful, nothing beats testing your site on actual devices. This can identify any issues that might not be apparent in a desktop testing environment.

- **Continuous testing and iteration**: Responsive design is not a set-and-forget process. Continuously test and iterate your breakpoints as new devices enter the market and as user behavior changes.

By following these best practices, you can ensure that your breakpoints are set in a way that truly benefits the user experience. Remember, the goal of responsive design is to create web experiences that are as functional and enjoyable on a mobile device as they are on a large desktop screen.

Next, we'll explore how to walk the line between adaptability and overcomplication.

Adaptability versus overcomplication

Adaptability in web design is about creating websites that can adjust gracefully to different screen sizes, resolutions, and user contexts. It's a crucial aspect of modern web development, ensuring that a site is accessible and provides an optimal user experience across various devices. However, there's a fine line between making a design adaptable and overcomplicating it. This section will focus on defining what adaptability means in the context of web design and how to achieve it effectively.

At its core, adaptability in web design means that a website responds to the environment in which it is viewed. This involves dynamically adjusting layouts, content, and functionalities based on the device's capabilities and the user's needs.

Here is a list of adaptability benefits:

- **Fluid layouts**: Adaptability is often achieved through fluid grid layouts. Unlike fixed layouts that might look great on one device but break on another, fluid layouts use relative units such as percentages, which allow elements to resize in proportion to the viewing area.

- **Responsive images and media**: Adaptable web design also involves using responsive images and media. This means that images and videos should change size and resolution based on the screen they are displayed on, to ensure they load quickly and look crisp on all devices.

- **CSS media queries**: Media queries are the cornerstone of adaptability, allowing designers to apply different styles based on certain conditions, such as screen width, height, and orientation. They enable the creation of multiple viewing experiences within a single design.

- **Progressive enhancement**: This approach focuses on building a functional core experience that works for everyone and then enhancing it for users with more capable browsers or devices. It ensures that the website is usable even on the most basic devices or in poor network conditions.

- **Accessibility considerations**: An adaptable website must also be accessible. This includes designing for keyboard navigation, screen readers, and ensuring that all interactive elements are easily accessible regardless of device.

- **Testing across devices and browsers**: Regular testing across different devices and browsers is essential for ensuring adaptability. It can identify any issues that might not be immediately apparent, such as touch target sizes on mobile devices or layout issues on different browsers.

While striving for adaptability, it's important to avoid overcomplication. This means not adding unnecessary breakpoints or overly complex navigation structures that could confuse users or make maintenance difficult.

In conclusion, adaptability in web design is about creating flexible, responsive, and accessible web experiences that cater to a diverse range of devices and user needs. By focusing on fluid layouts, responsive media, and progressive enhancement, while avoiding unnecessary complexity, you can ensure that your website is not only adaptable but also efficient and user-friendly.

Signs of overcomplication in responsive design

While adaptability in web design is essential for creating inclusive and accessible digital experiences, there's a risk of overcomplicating the design in an effort to accommodate every possible scenario. Overcomplication can lead to a range of issues, from poor performance to maintenance challenges. Recognizing the signs of overcomplication is crucial for ensuring that your responsive design remains effective and user-friendly.

Here are some signs of overcomplication:

- **Excessive use of media queries**: One of the primary indicators of overcomplication is the overuse of media queries. While media queries are fundamental to responsive design, relying on an excessive number can lead to bloated and hard-to-maintain CSS. Ideally, media queries should be used only when there's a clear need for them, based on major shifts in content layout or functionality.

- **Overly complex navigation systems**: Responsive design often requires different navigation approaches for different devices. However, if your navigation system becomes too complex or inconsistent across devices, it can confuse users. Overcomplication in this area might involve multi-level dropdowns, hidden menus that are difficult to access, or varying navigation patterns that lack coherence.

- **Too many breakpoints**: While breakpoints are essential in responsive design, having too many can be a sign of overcomplication. A website doesn't need to cater to every possible screen size. Instead, focus on the major categories of devices (mobile, tablet, and desktop) to simplify your design and reduce the likelihood of unexpected issues.

- **Performance issues**: Overcomplicated designs often lead to performance issues. This can manifest as slow loading times, especially on mobile devices. If your website is packed with high-resolution images, complex animations, and heavy scripts to accommodate various scenarios, it's likely to perform poorly on less capable devices or slower networks.

- **Difficulty in maintenance**: If updating or maintaining your website becomes a significant challenge due to the complex structure of your responsive design, it's a clear sign of overcomplication. Responsive design should aim to simplify updates, not complicate them. This is often a consequence of redundant or overly specific CSS rules and JavaScript functions.

- **Inconsistent user experience**: The user experience should be consistent and intuitive across all devices. Overcomplication can lead to a disjointed user experience, where elements behave unpredictably, or layouts change dramatically and unnecessarily between breakpoints.

- **Ignoring content hierarchy**: An effective responsive design maintains a clear content hierarchy across devices. Overcomplication might result in important content being pushed down or hidden in an attempt to make everything fit, regardless of its priority or relevance to the user.

In conclusion, while creating a responsive design that adapts to various screen sizes and devices, it's important to be mindful of these signs and the results of overcomplication. Striking the right balance between adaptability and simplicity will not only enhance the user experience but also ensure that your website is manageable and performs well across all platforms.

Balancing flexibility and simplicity

In RWD, achieving a balance between flexibility and simplicity is key to creating an effective and user-friendly experience. While it's important for a design to adapt to various screen sizes and devices, maintaining simplicity should be a guiding principle. This balance ensures that the website is both functional and accessible, without overwhelming the user or the developer.

Let's dive into how you can find the best balance between flexibility and simplicity:

- **Prioritize UX**: The primary goal of responsive design should always be to enhance the user experience. This involves ensuring that navigation is intuitive, content is easily accessible, and the overall layout is clean and uncluttered, regardless of the device being used.

- **Adopt a mobile-first approach**: Starting with a mobile-first design focuses you on the most essential elements of your site. This approach inherently leads to a simpler design, as it requires prioritizing content and functionalities that are crucial for small screens. These priorities can then be expanded upon for larger screens.

- **Use conditional loading wisely**: Conditional loading can be a powerful tool in responsive design, allowing different content to be loaded depending on the device. However, it's important to use this technique judiciously to avoid overloading the user with unnecessary information or complicating the backend logic excessively.

- **Minimize breakpoints**: While breakpoints are essential for responsive design, using too many can lead to complexity. Aim to identify key screen sizes where your content and layout need significant adjustments, rather than trying to cater to every possible device.

- **Streamline content and visuals**: Simplify your content and visual elements so they are effective across all devices. This means using graphics and text that are legible and engaging on both small and large screens and avoiding excessive decorative elements that don't add functional value.

- **Optimize navigation**: Navigation is a critical aspect of UX and should be as straightforward as possible. Simplify your site's navigation to ensure that users can easily find what they need, whether they are on a touch screen or using a mouse and keyboard.

- **Regular testing and user feedback**: Regularly test your design on various devices and gather user feedback. This process can help identify areas where the design may be too complex or not adaptable enough, allowing adjustments that better balance flexibility and simplicity.

- **Performance as a measure of simplicity**: Monitor your site's performance as an indicator of its simplicity. A well-optimized, fast-loading site is often a sign that the design has struck the right balance between being adaptable and not overly complex.

In conclusion, balancing flexibility with simplicity in responsive design is about making strategic choices that prioritize the user experience. By focusing on the essential elements, streamlining content and visuals, and minimizing unnecessary complexity, you can create a responsive design that is both adaptable and user-friendly.

Now, let's explore the implications of breakpoints in the real world.

Real-world implications of breakpoint choices

In the context of strategic breakpoints and adaptability, understanding the impact of breakpoints on user experience is vital. Breakpoints, the points in responsive design where the website content and layout adjust to accommodate different screen sizes, play a critical role in shaping how users interact with a website. Their strategic placement can greatly enhance user experience, while poorly considered breakpoints can lead to frustration and disengagement.

Here are some benefits of good breakpoint choices:

- **Ensuring seamless navigation and readability**: Effective breakpoints ensure that website navigation remains intuitive, and content remains readable across all devices. For instance, a menu that works well on a desktop might need to be collapsed into a hamburger icon on mobile devices. Similarly, text and images need to be re-sized or re-stacked to remain legible and visually appealing.

- **Adapting to user expectations**: Users' expectations depend on the device they are using. On mobile devices, they might expect faster access to information and simplified browsing, while on desktops, a more comprehensive view might be preferred. Properly placed breakpoints allow for these different user expectations to be met effectively.

- **Reducing cognitive load**: Good breakpoint implementation reduces the cognitive load on users. It prevents information overload on small screens by reorganizing or prioritizing content and functionalities based on device capabilities, making the site easier to navigate and understand.

- **Enhancing engagement**: Breakpoints that optimize content for various screen sizes can significantly increase user engagement. A well-designed responsive site keeps users interested and interacting with the content, regardless of their device.

- **Minimizing layout shifts**: Inconsistent or poorly planned breakpoints can lead to unexpected layout shifts as users switch between devices or orientations, negatively impacting the user experience. Strategic breakpoints ensure consistency in the user interface, providing a more stable and predictable interaction.

- **Improving accessibility**: Well-implemented breakpoints also enhance accessibility. They ensure that content is not just visually adaptable but also functional for users with varying abilities and those using assistive technologies on different devices.

- **Impact on user satisfaction and retention**: Ultimately, the way breakpoints are handled can significantly impact overall user satisfaction and retention. A responsive design that adapts smoothly to different screen sizes leaves users with a positive impression of the site, encouraging them to return.

In conclusion, the impact of breakpoints on user experience is profound. They are not just technical necessities but crucial elements in crafting a responsive website that meets users' needs and expectations. Thoughtfully implemented breakpoints contribute to a seamless, engaging, and accessible user experience, which is essential in today's diverse device landscape.

Breakpoints and website performance

Understanding the relationship between breakpoints and website performance is essential. Breakpoints, which define how a website's layout adjusts at certain screen sizes, have a direct impact on how a site performs across different devices. When strategically set, they can enhance a site's efficiency and speed, but if poorly managed, they can lead to performance issues.

Here's how breakpoints can improve website performance:

- **Optimizing load times**: Properly implemented breakpoints can optimize load times. By adjusting image sizes, layout structures, and content visibility at different breakpoints, the website can load only the necessary elements for each device, preventing the loading of heavy resources that are unnecessary for smaller screens.

- **Responsive resource delivery**: Breakpoints play a key role in determining how resources are delivered to different devices. By using techniques such as responsive image delivery, where images of different resolutions are served based on the device, breakpoints can significantly reduce bandwidth usage and improve the site's performance.

- **Balancing design complexity**: Breakpoints should be used to balance the complexity of the design with performance. Overly complex designs with too many breakpoints can lead to increased HTTP requests and CSS calculations, slowing down the site. A strategic approach to breakpoints can maintain an optimal balance.

- **Enhancing user-perceived performance**: From a user's perspective, performance is often judged by how quickly the site becomes usable. Effective breakpoint management can enhance perceived performance by ensuring that the most important content is loaded quickly and is appropriately laid out for immediate interaction.

- **Impact on server load**: With adaptive design techniques linked to breakpoints, server-side components can deliver optimized content based on the user's device. This adaptive content strategy can reduce server load and improve overall site performance.

- **Breakpoints and JavaScript performance**: Implementing breakpoints often involves JavaScript, especially for dynamic content reorganization. It's important to ensure that these JavaScript implementations do not become performance bottlenecks, particularly on devices with lower processing power.

- **Testing and optimization**: Regular performance testing across different breakpoints is crucial. This should include evaluating how different elements load and interact at various breakpoints, ensuring that the site remains performant at all screen sizes.

In conclusion, the way breakpoints are implemented can have a significant impact on website performance. They are not just about visual adaptation but also play a key role in how efficiently and effectively a website operates on different devices. Thoughtful consideration of breakpoints in the context of performance is essential for creating a responsive, fast, and user-friendly website.

Case studies — Successes and pitfalls in breakpoint implementation

While exploring the real-world implications of breakpoint choices, examining case studies provides valuable insights into the successes and pitfalls associated with breakpoint implementation. These real-life examples can clarify how strategic or misguided breakpoint choices can significantly impact a website's functionality and user experience.

Let's begin:

- **Case Study 1: E-commerce success through strategic breakpoints**

  A prominent e-commerce platform redesigned its website with a focus on responsive breakpoints. They implemented breakpoints not just based on device widths but also considered the type of content displayed. This approach resulted in a significant increase in mobile conversion rates. Key success factors included optimizing product images and call-to-actions to appear prominently on smaller screens and reorganizing content for better readability. This case study demonstrates the importance of aligning breakpoints with user needs and content strategy.

- **Case Study 2: News website and overcomplication**

  A well-known news portal encountered issues when it overcomplicated its breakpoint strategy. It implemented many breakpoints in an attempt to tailor the experience for every possible screen size. This led to a cluttered CSS, increased load times, and maintenance difficulties. User engagement dropped, especially on mobile devices, due to slow performance and a confusing layout. This case highlights the pitfalls of overcomplicated breakpoint implementation and underscores the need for a balanced approach.

- **Case Study 3: Adaptive design in a tech blog**

  A technology blog successfully utilized adaptive design techniques in conjunction with breakpoints. It designed its content to adapt dynamically at each breakpoint, enhancing readability and engagement across devices. The site used larger images and a different layout for desktop users, while mobile users received a streamlined experience with compressed images and a simplified layout. The result was an increase in both traffic and average session duration. This example illustrates the effectiveness of combining adaptive content strategies with well-placed breakpoints.

- **Case Study 4: Small business website with insufficient breakpoints**

  A local restaurant's website failed to implement sufficient breakpoints, resulting in a subpar mobile experience. The site worked well on desktops but lacked responsiveness on smartphones, with some content becoming inaccessible or poorly formatted. This oversight led to a decrease in online reservations from mobile users. The restaurant later revised its approach, implementing additional breakpoints to optimize the mobile experience, which led to improved customer interaction and increased reservations.

Let's look at the key takeaways from the case studies.

The e-commerce platform and tech blog exemplify the benefits of thoughtfully implemented breakpoints, tailored to content and user behavior. On the other hand, the news portal and restaurant website demonstrate the consequences of either overcomplicating or oversimplifying breakpoint strategies. These case studies underline the necessity of a balanced and user-centric approach in breakpoint implementation for successful RWD.

In conclusion, these case studies emphasize the impact of strategic breakpoint choices on a website's success. Learning from both the triumphs and mistakes of others can help designers and developers make informed decisions that enhance user experience and website performance.

Summary

As we wrap this chapter up, let's reflect on the essential insights and skills that have been imparted. We've navigated through the current trends in device usage, understanding how they shape the need for RWD. You've learned the critical methodology behind selecting effective breakpoints, a skill fundamental to crafting websites that respond seamlessly to varying screen sizes and devices.

We've also delved into the delicate balance between adaptability and overcomplexity in design. This understanding is vital to avoid overwhelming users and complicating website maintenance, ensuring a design is as functional as it is aesthetically pleasing. Additionally, the chapter highlighted the real-world implications of breakpoint decisions, emphasizing how these choices directly impact user experience and website performance.

The skills and knowledge gained here are invaluable for any web professional. They enable you to create user-friendly, efficient, and responsive websites, meeting the diverse needs of today's digital audience.

In the next chapter, we'll build upon these foundations, focusing on how to effectively guide users through your website to maximize conversions. This next step is a natural progression, as understanding breakpoints and adaptability sets the stage for creating intuitive navigation and clear conversion paths, essential elements in driving user engagement and achieving business goals.

7

User Navigation and Conversion Pathways

Welcome to *Chapter 7*, a comprehensive guide designed to deepen your understanding of creating effective, user-friendly navigation systems that cater to diverse user needs across various devices. In today's digital landscape, where user interaction spans from desktops to mobile devices, mastering the art of intuitive navigation is crucial for any website's success.

In this chapter, we will explore the principles behind user-friendly navigation. This includes understanding how to structure your website's navigation to make it intuitive and straightforward so that users can find what they need with ease. We'll delve into the world of mobile navigation, recognizing that users on mobile devices have different expectations and limitations compared to desktop users.

You'll also learn how conversion paths vary across devices. This knowledge is essential for tailoring user experiences that lead to higher engagement and conversion rates. Whether a user is browsing on a large desktop screen or a compact smartphone display, the path to conversion should be clear and accessible.

Additionally, this chapter addresses common navigational design flaws. By identifying and rectifying these issues, you can enhance the overall usability of your website, improving user satisfaction and performance metrics.

By the end of this chapter, you'll have gained valuable skills in crafting navigation systems that are not only aesthetically pleasing but also functionally robust, meeting the needs of today's diverse internet users. We aim to equip you with the knowledge to create navigation that guides users smoothly from their entry point to conversion, regardless of their device of choice.

The following topics will be covered in this chapter:

- The principles of intuitive navigation
- Mobile menus and user expectations
- The path to conversion on varied devices
- Overcoming navigation pitfalls

The principles of intuitive navigation

In the world of web design, establishing a clear navigation hierarchy is paramount. This isn't just about organizing content; it's about crafting an intuitive pathway that guides users through your website seamlessly. A logical and clear structure not only enhances usability but also enriches the user experience, playing a crucial role in the effectiveness of conversion pathways.

The essence of intuitive navigation starts with organizing the website's content in a way that mirrors the user's thought process. Imagine your website as a map; the main categories are the big, bold cities, easily spotted and recognized, while subcategories are like smaller towns, important but branching out from the main routes. Each element should be distinct and self-explanatory, leading users naturally from one point to the next.

Prioritization is key in this structure. The most vital information or features should take center stage and be prominently displayed and easily accessible. This could be your primary services, products, or calls to action — the pillars of your website. Less critical information, while still accessible, shouldn't overshadow these primary elements.

Visual cues play a significant role in reinforcing this hierarchy. The use of size, color, typography, and spacing can direct attention and suggest importance. Larger, bolder items naturally draw the eye and should be used for primary navigation elements. Using contrasting colors effectively can highlight critical features or calls to action, guiding the user's journey through the website.

Consistency in this hierarchy across different pages is a subtle yet powerful tool. It ensures that once users become familiar with your navigation layout, they can confidently navigate the rest of the site. This reliability reduces the effort needed to understand the website, making navigation almost second nature.

In the age of responsive design, this navigation hierarchy must elegantly adapt to various screen sizes. On mobile devices, what was a horizontal menu bar might become a compact hamburger menu, but the core principle remains the same — clear, accessible navigation.

However, a navigation hierarchy should never be set in stone. It's an evolving aspect of your website, requiring regular refinement and adaptation based on user feedback and interaction data. Through analytics and user testing, you can gain insights into how effectively your navigation guides users and makes necessary adjustments, ensuring that your website not only meets but anticipates user needs.

In essence, a well-structured navigation hierarchy is the backbone of a user-friendly website. It guides users fluidly, supports your site's goals, and enhances the overall user experience. By prioritizing content logically, maintaining consistency, and adapting to different devices, you create more than just a pathway through your website — you create a journey that users are eager to take.

Next, we'll explore how to design for usability across devices.

Designing for usability across devices

In the digital landscape, where user interaction spans a variety of devices, designing for usability across these diverse platforms is a cornerstone of effective web design. The challenge lies in creating a navigation experience that is intuitive and consistent, whether a user is clicking with a mouse, tapping on a touchscreen, or using any other input method.

The fundamental step in designing for cross-device usability is acknowledging the different interaction models of each device. Desktop users often rely on precise mouse clicks and enjoy the luxury of larger screens, while mobile users navigate with taps and swipes on smaller displays. Tablet users might find themselves in a hybrid situation, using both touch and keyboard inputs. Catering to these varying interactions requires a thoughtful approach to how navigation elements are structured and behave.

Responsive design extends beyond simply resizing content to fit different screen sizes. It encompasses rethinking navigation structures. On desktops, you might have the space for expansive menus and detailed navigation options. However, these need to be compacted or restructured for mobile screens, often through collapsible menus or simplified icons that don't compromise the discoverability of key content.

Despite these adaptations, maintaining a level of consistency is crucial. Users should feel a sense of familiarity when they switch between devices. This can be achieved by keeping key navigation elements consistent in terms of appearance and location, ensuring that users don't feel lost when they change devices.

For mobile devices, in particular, ensuring that all navigation elements are touch-friendly is essential. This includes making buttons and links large enough that they can easily be tapped and spacing them sufficiently to prevent accidental taps. It's not just about scaling down the desktop version; it's about reimagining navigation for touch interaction.

Simplicity in navigation design enhances usability. Avoiding overly complex or multi-layered navigation menus can prevent users from becoming overwhelmed or lost. Clear, concise labels and intuitive icons can significantly aid in guiding users through your website, regardless of the device they're using.

Lastly, testing your navigation on various real devices is as important as the design process itself. What looks good and works well on one device may not translate effectively to another. Regular testing ensures that your navigation design truly meets the cross-device usability it aims for.

In summary, designing for usability across devices is not just a technical challenge; it's about deeply understanding and catering to the user's context and needs. By focusing on responsive design, maintaining consistency, optimizing for touch interactions, and embracing simplicity, you can create a navigation experience that is intuitive and satisfying, no matter how your users access your website.

Now, let's look to incorporate user feedback.

Incorporating user feedback and behavior into design

In the journey of creating intuitive and effective web navigation, incorporating user feedback and understanding user behavior is pivotal. This process not only enhances the design but also aligns it more closely with the needs and expectations of the end users. It's a crucial step in ensuring that the navigation system of a website is not just theoretically sound but practically efficient and user-friendly.

Direct feedback from users is an invaluable resource in refining web navigation. This feedback can be gathered through various means, such as surveys, user interviews, and feedback forms. It provides direct insights into what users like and dislike, as well as what they find confusing about the website's navigation. Incorporating this feedback allows designers to make informed adjustments that cater specifically to their user base.

Web analytics tools provide a wealth of data on how users interact with a site. By analyzing metrics such as click-through rates, navigation paths, and time spent on different pages, designers can gain a deeper understanding of user behavior. This data helps in identifying patterns and trends that can inform navigation improvements. For example, frequently visited pages might need to be made more accessible, or complex navigation paths might need to be simplified.

Heatmaps are another powerful tool for understanding user navigation behavior. They provide a visual representation of where users click, move, and scroll on a site. This visual data can highlight which navigation elements are effective and which are being overlooked or causing confusion.

A/B testing involves comparing two versions of a web page to see which performs better. By implementing slight variations in the navigation layout or design, and measuring how these changes impact user behavior, designers can iteratively improve the navigation. This method ensures that changes are data-driven and have a positive impact on user experience.

Creating user personas and mapping out user journeys can also play a significant role in navigation design. This involves understanding the different types of users who visit the website and their objectives. Designing navigation with these personas and journeys in mind ensures that it caters to the specific needs and preferences of different user segments.

Incorporating user feedback should also include ensuring that navigation is accessible to all users, including those with disabilities. This means considering aspects such as keyboard navigation, screen reader compatibility, and clear labeling.

In conclusion, incorporating user feedback and behavior into navigation design is not just beneficial; it's essential for creating a user-centric website. This approach ensures that the navigation system is grounded in real-world usage, making it intuitive, efficient, and accessible to all users. By continually listening to users and observing their behavior, designers can create navigation that truly resonates with and serves the needs of their audience.

Now, let's dive into the fun world of mobile menus.

Mobile menus and user expectations

Adapting to mobile user behaviors is a crucial aspect of designing effective mobile menus. The shift to mobile browsing has brought with it distinct user behaviors and expectations that significantly differ from desktop browsing. Understanding and accommodating these behaviors is key to ensuring a smooth, intuitive mobile user experience.

Mobile browsing typically happens on the go, implying that users often seek information quickly and with minimal effort. This context demands a navigation system that is streamlined and straightforward. Unlike desktop users, who might be willing to navigate multiple layers of a menu, mobile users generally prefer to access what they need in as few taps as possible.

Given the limited screen space, it's essential to simplify navigation elements. This involves prioritizing the most important menu items and possibly reducing the number of visible elements. The challenge is to balance providing enough navigational options and keeping the interface uncluttered.

Mobile devices are predominantly touch-based, which necessitates larger, easily tappable menu items. This is crucial not only for enhancing usability but also for ensuring accessibility. Design elements such as buttons, links, and menu toggles must be sized and spaced in a way that they can be easily tapped without the risk of mis-clicks.

Essential features and information should be readily accessible. For instance, e-commerce sites might prioritize access to search functions, cart, and user account sections. This prioritization should be based on an understanding of what mobile users most commonly seek when they visit your site.

Familiarity breeds usability in mobile navigation. Utilizing common mobile menu patterns, such as the "hamburger" menu, can be beneficial as most users are already accustomed to these conventions. However, it's also important to critically evaluate if such patterns are the best fit for your site's specific user needs.

While striving for consistency in the navigation experience across devices, it's crucial to adapt to the unique aspects of mobile usage. The goal is to create a mobile version of the menu that feels familiar to the desktop version but is optimized for mobile interactions.

Mobile user behaviors and expectations are constantly evolving. Regular testing and updates are necessary to ensure that your mobile navigation remains effective and aligned with current user trends. This includes gathering user feedback, conducting usability tests, and keeping abreast of new mobile interaction patterns.

In summary, adapting to mobile user behaviors in menu design is about understanding the unique context and limitations of mobile devices and responding with a navigation experience that is intuitive, accessible, and user-friendly. By focusing on simplicity, touch-friendly design, and prioritization of key elements, designers can create mobile menus that effectively meet the needs of their mobile audience.

Now, let's look at how we can design effective menus.

Designing effective hamburger menus and alternatives

Designing mobile menus, particularly the widespread "hamburger" menu, is a critical consideration. While the hamburger menu has become a staple in mobile design due to its space-saving efficiency, it's equally important to explore its effectiveness and consider viable alternatives. This balance is key to meeting diverse user expectations and enhancing mobile usability.

The hamburger menu, characterized by its three horizontal lines resembling a hamburger, is a popular choice for mobile sites due to its minimalist appearance and ability to declutter the interface. However, it's not without its drawbacks. One significant issue is discoverability; users may not always realize that this icon houses the main navigation. When employing a hamburger menu, it's essential to make it prominent and intuitive, perhaps with accompanying text such as "Menu" for clarity.

When utilizing a hamburger menu, the design should focus on intuitive navigation. The menu should open in a way that is natural for users, typically full-screen or as a slide-out panel. The contents should be clearly organized, with important items at the top. If the menu is extensive, consider grouping items into categories or using accordions to save space and reduce scrolling.

Providing interactive feedback when users open the hamburger menu can enhance the experience. This could be a subtle animation of the icon transforming into a close ("X") symbol, indicating to users that they can revert to the main page by pressing it.

Alternatives to the hamburger menu include tab bars or bottom navigation bars, which are excellent for providing immediate access to four to five of the most important navigation destinations. Another option is the priority+ pattern, where the most important items are visible, and the rest are tucked behind a "More" option.

When designing alternatives, prioritize content based on user needs and the goals of your site. For instance, in a bottom navigation bar, include icons alongside labels for clarity and faster recognition. Each alternative should be tested for usability, ensuring that it aligns with the typical user behavior and preferences on your site.

It's crucial to adapt the navigation pattern based on observed user behaviors and preferences. Regularly analyze user interaction data and gather feedback to understand which navigation style best suits your audience.

In conclusion, whether you opt for a hamburger menu or its alternatives, the key lies in designing for ease of use, intuitive navigation, and aligning with user expectations. Effective mobile menu design should facilitate quick and easy access to the most important sections of your site, enhancing the overall mobile user experience. Regular testing and adaptation based on user feedback are essential to ensure the chosen navigation method remains effective and user-friendly.

Keep in mind that we always need to ensure our mobile menus are clear and accessible. Let's understand how we can go about ensuring that.

Ensuring clarity and accessibility in mobile menus

In the digital realm, ensuring clarity and accessibility in mobile menus is pivotal for creating an inclusive and user-friendly mobile experience. Since mobile devices have become the primary means of accessing the internet for many users, the need for clear, accessible navigation cannot be overstated. It's about crafting a mobile menu that not only guides users effortlessly but is also accessible to everyone, including those with disabilities.

Clarity in mobile menu design is about making navigation intuitive and straightforward. This involves using clear, descriptive labels for menu items and avoiding technical jargon that might confuse users. Icons can be used alongside text to provide visual cues, but they should be familiar and universally understood. Remember, a menu that is easy to understand is a menu that is easy to use.

Organizing the menu logically is critical. Group related items together and order them according to their importance and relevance. The most frequently accessed items should be placed where users can find them easily, usually at the top of the menu. A well-structured menu helps users find what they need quickly, enhancing their overall experience on your site.

Given the touch-based nature of mobile devices, menus must be designed with touch usage in mind. This includes making menu items large enough to be easily tapped without accidental activations. Adequate spacing between items is also important to prevent mis-selections, especially for users with motor impairments.

For users who rely on keyboards or screen readers, ensure that your mobile menu is fully navigable using these tools. This includes providing focus indicators for active elements and ensuring that all parts of the menu can be accessed and activated via keyboard commands. Screen reader users should be able to hear a clear and accurate description of each menu item.

Visual design elements such as contrast and font size play a significant role in menu clarity. Ensure high contrast between text and background colors for readability. Fonts should be large enough to be easily read on small screens. Also, consider providing options for users to adjust the text size as per their needs.

One of the most effective ways to ensure your mobile menu's clarity and accessibility is by testing it with real users, including people with disabilities. This can provide invaluable insights into how different users interact with your menu and highlight areas for improvement.

Continuously adapt your mobile menu based on user feedback and usage data. User needs and behaviors can change, and your menu should evolve accordingly to remain effective and accessible.

In summary, a clear and accessible mobile menu is not just an enhancement; it's a necessity for creating an inclusive web experience. By focusing on clarity, logical structure, touch accessibility, and compatibility with assistive technologies, you can design a mobile menu that meets the diverse needs of all your users. Regular testing and adaptation are key to maintaining its effectiveness and ensuring that your website remains welcoming and accessible to everyone.

Now, let's explore quick access and how that relates to menus.

Optimizing menu content for quick access

Optimizing menu content for quick access on mobile devices is essential. Mobile users often seek immediate information and value speed and efficiency. Therefore, a well-optimized mobile menu is key to providing a positive user experience, enhancing usability, and ultimately guiding users toward conversion pathways more effectively.

The core of optimizing menu content is prioritization. Analyze user interaction data to identify the most frequently accessed sections of your site. These elements should be the most easily accessible within your mobile menu. Prioritizing content based on user behavior ensures that the menu aligns with users' needs and expectations.

On mobile devices, screen real estate is limited. Simplifying your menu so that it only includes the most essential items is crucial. Each menu item should be clear and concise, reducing the cognitive load on the user. Avoid technical jargon and opt for simple, direct language that users can quickly understand.

A responsive mobile menu should adapt not just in size but also in content based on the device being used. For example, options that are critical for mobile users, such as a "Call now" button, should be more prominent than on a desktop.

Using icons in your mobile menu can aid in quick recognition of menu items, saving space and adding to the visual appeal. However, ensure that the icons are intuitive and complemented by text labels to avoid any ambiguity.

Aim to minimize the number of taps a user must make to reach their desired page. This might involve reducing the depth of your menu. Consider features such as expandable menus or a search bar within the menu to help users quickly navigate to deeper sections of the site.

Continually test your mobile menu with real users to gather feedback on its usability and efficiency. User testing can reveal insights that are not immediately apparent and help refine the menu for quicker access.

Ensure that your mobile menu loads quickly and functions smoothly. Delays in loading or laggy interactions can frustrate users and negatively impact their experience.

Accessibility should not be overlooked in pursuit of speed. Ensure that your mobile menu is accessible to all users, including those with disabilities. This includes readable text sizes, sufficient contrast, and support for screen readers.

In conclusion, optimizing a mobile menu for quick access involves a thoughtful balance of content prioritization, simplicity, and usability considerations. A well-optimized mobile menu not only enhances the overall user experience but also facilitates smoother navigation to conversion-relevant areas of your site. Regular testing and iteration based on user feedback and behavior are crucial in maintaining a menu that effectively meets the needs of your mobile audience.

The path to conversion on varied devices

We also need to think about optimizing menu content for quick access on mobile devices. With the majority of internet users accessing the web through mobile devices, the need for efficient, easily navigable menus is more critical than ever. The goal is to provide a seamless and quick navigation experience that allows users to find what they need with minimal taps and in the shortest time possible.

The first step in optimizing your mobile menu is to prioritize the most important information. This involves identifying what your users are most likely seeking when they visit your site. For example, an e-commerce site might prioritize product categories, a search bar, and a shopping cart, while a news site might focus on trending topics and categories.

A cluttered menu can overwhelm and confuse users, leading to a poor experience. Simplify your menu by limiting the number of options. This doesn't mean omitting essential information but rather organizing it more efficiently, such as using dropdowns or accordion menus for less critical items. The idea is to present users with a clean, straightforward menu where choices are clear and easy to navigate.

In a space-constrained mobile environment, the use of icons alongside or in place of text can be a great way to save space while maintaining clarity. However, it's crucial to choose icons that are universally recognizable and pair them with clear, concise labels. This combination ensures that users can quickly understand their options, even at a glance.

For sites with extensive content, integrating a prominent and efficient search function in the mobile menu can significantly enhance the user experience. A well-placed search bar allows users to bypass traditional navigation paths and directly find the content they need.

Responsive menus that adapt to the changing needs and behaviors of users can significantly improve access speed. This might include dynamically changing menus based on user interactions, location, time of day, or most frequently accessed content, offering a personalized and efficient navigation experience.

User expectations and behaviors evolve, and so should your mobile menu. Regular user testing, including A/B testing of different menu layouts and structures, can provide invaluable insights. Use this feedback to iterate and refine your menu, ensuring that it continually meets the needs of your users.

The efficiency of a mobile menu isn't just about its layout and content but also how quickly it loads. Ensure that your menu is optimized for speed; a slow-loading menu can negate all efforts put into organizing and simplifying it.

In summary, optimizing a mobile menu for quick access is about understanding and prioritizing user needs, simplifying choices, and ensuring clarity and speed. It's a process of continual refinement and adaptation to user feedback, aimed at providing the best possible navigation experience on mobile devices. By focusing on these key aspects, you can create a mobile menu that not only meets but exceeds user expectations.

Now, let's explore **calls to action (CTAs)**.

Optimizing CTA elements for different screens

A key aspect that stands out is the optimization of CTA elements for different screens. CTAs play a pivotal role in guiding users toward conversion, be it making a purchase, signing up for a newsletter, or engaging with content. The challenge lies in ensuring these CTAs are equally effective across the varied screen sizes and devices that users employ to interact with websites.

Here are some ways to optimize a CTA:

- **Adapting the CTA's size and placement**: On smaller screens, such as smartphones, space is at a premium. CTAs need to be prominent enough to be noticed but not so large as to overwhelm the content. The placement also matters — a CTA at the bottom of the screen is easier to tap with a thumb, improving the usability of touch devices. On larger screens, such as desktops, you have more leeway to integrate CTAs into your design without space constraints.

- **Clear and compelling messaging**: The message on a CTA needs to be clear and compelling, regardless of the screen size. On mobile devices, where users are likely to scan content quickly, a concise, action-oriented CTA can be more effective. On desktops, you have the space to include more detailed messages if necessary, but the principle of clarity should still guide your copy.

- **Contrasting colors for visibility**: The use of contrasting colors is crucial in making CTAs stand out. This is especially important on smaller screens, where every element competes for the user's attention. A CTA that pops out visually can draw the user's eye more effectively, thereby increasing the likelihood of a click or tap.

- **Responsive button design**: The design of the CTA button itself should be responsive. It needs to be large enough to be easily tapped on a mobile device — a critical consideration for touch interfaces. At the same time, it should not appear disproportionately large or out of place on larger screens.

- **Testing across devices**: Regularly testing CTAs across different devices is vital. What works on a desktop might not translate well to a mobile interface, and vice versa. A/B testing different versions of CTAs can help you identify the most effective design, placement, and messaging for various screen sizes.

- **Consistency across devices**: While adaptation to different screens is necessary, maintaining a degree of consistency in the appearance and feel of CTAs across devices is important. This consistency helps in building a coherent brand image and user experience.

- **Loading speed and interactivity**: The loading speed of CTA elements, particularly on mobile devices, can impact their effectiveness. Additionally, incorporating a degree of interactivity, such as a subtle animation upon a hover (on desktops) or a touch (on mobile devices), can make CTAs more engaging.

In conclusion, optimizing CTAs for different screens is a multifaceted process involving thoughtful adaptation of design elements, clear messaging, and continuous testing. By ensuring that these crucial elements are effectively tailored for various devices, websites can maximize their potential for user engagement and conversion, crucial for any digital strategy.

Next, we'll explore how to make the checkout process simple.

Streamlining the checkout process for mobile users

A critical focus should be on streamlining the checkout process for mobile users. With the increasing prevalence of mobile commerce, ensuring a smooth, fast, and user-friendly checkout process on mobile devices is essential for driving conversions and reducing cart abandonment rates.

The first step in optimizing the mobile checkout process is simplifying the interface. This means minimizing the number of steps and form fields users must navigate. Essential information should be prioritized, and any non-critical fields should be eliminated or made optional. A clean, uncluttered interface reduces confusion and helps keep users focused on completing their purchase.

On mobile devices, filling out forms can be tedious. Designing efficient, easy-to-use forms is crucial. This includes large, clickable areas for form fields, clear labeling, and using the appropriate keyboard for text entry (numeric keyboards for phone numbers, for example). Implementing features such as auto-fill and stored user information can significantly speed up the process.

Adding progress indicators is a subtle yet powerful way to guide users through the checkout process. Knowing how many steps are left in the process can reduce frustration and give users a sense of control and progress, which is especially important in a reduced mobile screen space.

Offering a variety of payment options caters to a broader range of users. Integrating mobile-friendly payment solutions such as Apple Pay, Google Wallet, or PayPal, which allow for one-tap payments, can make the checkout process faster and more convenient for mobile users.

Clear error messages and assistance within the checkout process can prevent user drop-off. If a user inputs information incorrectly, immediate, clear feedback on what needs to be corrected helps prevent confusion and frustration.

On mobile devices, where screen space is limited, prominently displaying security badges and trust signals without overcrowding the interface can reassure users, especially during the payment process. This helps build confidence in the transaction and reduces anxiety about sharing personal and payment information.

The checkout process needs to be responsive and fast-loading on mobile devices. Slow load times can be a major deterrent for users completing a purchase. Ensuring that the checkout process is optimized for performance is crucial.

Continuously testing and optimizing the mobile checkout process based on user feedback and behavior is essential. This includes monitoring conversion rates and cart abandonment rates, as well as conducting user testing to identify areas for improvement.

In conclusion, streamlining the checkout process for mobile users is about creating an efficient, secure, and user-friendly experience. By simplifying the interface, optimizing form fields, providing multiple payment options, and ensuring fast performance, businesses can significantly enhance the mobile checkout experience, leading to increased conversions and customer satisfaction.

Next, we'll cover conversation pathways.

Balancing information and simplicity in conversion pathways

An important aspect to consider is the balance between providing sufficient information and maintaining simplicity in conversion pathways, especially when dealing with varied devices. This balance is crucial in designing an effective user journey that leads to conversions without overwhelming or confusing the user.

Here are some ways to find balance:

- **Understanding user needs**: The key to balancing information and simplicity lies in understanding what users need to make a decision about. This understanding should be based on user behavior and analytics. For instance, while detailed product descriptions might be essential on an e-commerce site, a service-oriented site might benefit more from clear, concise CTAs.

- **Clarity in CTAs**: Clear and direct CTAs are essential for guiding users toward conversion. On different devices, these CTAs must stand out and be easily actionable. This could mean larger, more prominent buttons on mobile screens and more detailed, descriptive buttons on larger desktop screens.

- **Streamlining content**: While it's important to provide all the necessary information, the way it's presented should be streamlined. Avoid cluttering pages with excessive text or images, especially on mobile devices. Use expandable content sections such as accordions or tabs to organize information efficiently, allowing users to access it without overwhelming them initially.

- **Visual hierarchy**: A well-thought-out visual hierarchy helps maintain simplicity while providing information. Key elements that contribute to conversions should be more prominent. This hierarchy guides users' eyes to what's most important, whether it's product features, benefits, or the checkout button.

- **Responsive design**: Ensure that your design is responsive, not just in terms of its layout but also in terms of content presentation. Information that is easily digestible on a desktop might need to be presented differently on a mobile device. This could involve altering the content's layout, size, or even the amount of information based on the device being used.

- **User testing for feedback:** Regular user testing across various devices can provide insights into how effectively your site balances information and simplicity. Use this feedback to make adjustments, aiming for a design that efficiently guides users to conversion without unnecessary complexity.

- **Minimizing distractions:** In your pathway to conversion, minimize elements that can distract users from the main goal. This includes unnecessary links, excessive promotional messages, or intrusive popups, especially on smaller screens where space is limited.

- **Performance considerations:** The loading time of your pages, especially those critical to the conversion pathway, should be optimized. A fast-loading page ensures that users can access the information they need swiftly, contributing to a seamless experience.

Balancing information and simplicity in conversion pathways is a dynamic process that requires continuous attention and refinement. It's about presenting the right amount of information in a clear, accessible manner, adapted to the unique characteristics and limitations of various devices. By focusing on user needs, clarity in CTAs, streamlined content, and a strong visual hierarchy, the path to conversion can be made intuitive and effective, irrespective of the device used.

In the next section, we'll explore some navigation pitfalls and how to overcome them.

Overcoming navigation pitfalls

A critical step toward creating effective navigation is identifying common navigation errors. These errors, often subtle, can significantly impede the user experience and obstruct the path to conversion. Being aware of these pitfalls is the first step in creating a more intuitive and user-friendly navigation system, particularly in the context of varied devices.

Here's a list of the most common pitfalls:

- **Overly complex menus:** One of the most prevalent issues in web navigation is complexity. Menus that are overcrowded with options can overwhelm users, making it difficult for them to locate what they're seeking. This problem is exacerbated on mobile devices, where screen real estate is limited. Simplifying menus, categorizing items logically, and prioritizing content based on user needs are key to resolving this.

- **Inconsistent navigation structure:** Consistency in navigation structure across a website provides a sense of familiarity and ease for users. When navigation elements shift drastically from page to page, it disrupts the user's learning curve and increases the cognitive effort required to interact with the site. Consistency in the placement, style, and function of navigation elements is essential for a coherent user experience.

- **Hidden or obscure navigation elements:** Hidden navigation elements, such as those in hamburger menus or behind ambiguous icons, can sometimes lead to user frustration. While these designs can be effective in saving space, they should be implemented thoughtfully, ensuring that users can easily find and access these elements. Clarity and visibility are crucial, especially for key navigational items.

- **Lack of responsive design**: In today's multi-device world, a navigation system that fails to adapt responsively to different screen sizes can significantly degrade the user experience. A menu that works well on a desktop might be unusable on a smartphone. Ensuring that navigation elements are responsive and adaptable to various screen sizes is fundamental.

- **Ignoring user feedback**: Neglecting user feedback is a common pitfall. User feedback is a direct line to understanding the strengths and weaknesses of your site's navigation. Regularly soliciting and incorporating this feedback can provide valuable insights into how navigation can be improved.

- **Poorly designed drop-down menus**: Drop-down menus can be challenging, particularly if they are too long, too deep, or not organized intuitively. Users should be able to navigate these menus easily without feeling lost or overwhelmed. Consideration of touch interfaces on mobile devices is also essential.

- **Neglecting accessibility**: Accessibility in navigation is not just a legal requirement but a moral and practical one. Navigation should be designed with all users in mind, including those with disabilities. This includes keyboard navigability, screen reader compatibility, and clear, descriptive labeling.

Identifying and addressing these common navigation errors is essential in creating a smooth and intuitive user journey. By focusing on simplifying menus, maintaining consistency, ensuring visibility and responsiveness, listening to user feedback, designing dropdowns thoughtfully, and adhering to accessibility standards, you can enhance the usability of your site and facilitate a clearer path to conversion.

Next, we'll talk about some common responsive navigation challenges and their solutions.

Responsive navigation challenges and their solutions

Addressing the challenges of responsive navigation and implementing effective solutions is crucial for a seamless user experience across various devices. As users increasingly access content on a diverse range of screens, from smartphones to large desktop monitors, navigation must adapt fluidly. This adaptation, however, comes with its own set of challenges.

Let's consider some of these challenges:

- **Maintaining consistency across devices**: A primary challenge in responsive navigation is ensuring consistency in the user experience across different devices. Menus, links, and buttons must function predictably, regardless of the screen size or device. The solution lies in adopting a mobile-first approach, which emphasizes simplicity and scalability. Start by designing for smaller screens, where space is limited, and then expand the design for larger screens, adding elements as necessary without compromising the core experience.

- **Touch versus click interfaces**: Navigation designed for mouse clicks may not translate well to touch interfaces. The solution is to design with touch in mind from the outset. This includes making buttons and links large enough for comfortable tapping and spacing them to prevent accidental touches. Additionally, hover-based interactions should be rethought for touchscreens as they don't support hover states.

- **Hidden menus on mobile devices**: While hidden menus, such as hamburger menus, save space on mobile devices, they can also hide important navigation elements, potentially reducing discoverability. An effective solution is to complement hidden menus with visible key navigation options. For instance, the most important sections can be displayed as tabs, while additional options can be housed within the expandable menu.

- **Varied user navigation patterns**: Users navigate differently on mobile devices compared to desktops. On mobile, navigation should facilitate quick and direct access to information as users often seek immediate results. Implementing features such as a sticky menu bar, search functionality, and clear CTAs can help users find what they need without unnecessary navigation.

- **Load times and performance**: Responsive designs, particularly those with complex navigation structures, can suffer from longer load times. To combat this, optimize images and media, leverage caching, and minimize the use of heavy scripts and animations. The goal is to strike a balance between a visually appealing navigation design and efficient performance.

- **Screen orientation variability**: The shift between portrait and landscape modes on mobile devices can affect navigation layout. Responsive design should accommodate both orientations seamlessly. This can involve reflowing content, adjusting menu layouts, and ensuring that key navigation elements remain accessible, regardless of how the device is held.

In conclusion, overcoming the challenges of responsive navigation requires a thoughtful balance between functionality, aesthetics, and performance. By focusing on user-centric solutions, such as maintaining consistency across devices, designing for touch interfaces, optimizing key navigation elements for visibility, and ensuring fast load times, you can create a navigation system that effectively adapts to the diverse needs of today's digital audience.

Next up, we'll break down discoverability and accessibility.

Enhancing discoverability and accessibility

Enhancing discoverability and accessibility in web navigation stands as a critical aspect of designing user-friendly and inclusive digital experiences. Discoverability ensures that users can easily find the information they need, while accessibility guarantees that this is possible for all users, including those with disabilities. Addressing these aspects effectively can significantly improve the usability and reach of a website.

Here are some strategies you can use:

- **Improving discoverability**: Discoverability is about making key information and navigation elements easily findable. This can be enhanced by using clear, descriptive labels for navigation items while avoiding vague or jargon-filled terms. For sites with extensive content, incorporating a prominently placed search function can be invaluable. This feature allows users to bypass traditional navigation routes and quickly find specific information.

- **Logical navigation structure**: Organizing navigation logically and predictably is essential for discoverability. Users should be able to anticipate where to find information based on common web conventions and the logical grouping of items. This might include placing about us, contact information, or support sections in familiar locations, such as the website's footer.

- **Accessibility in navigation design**: Making navigation accessible means ensuring that it can be used by everyone, including people with disabilities. This includes providing keyboard navigability for users who cannot use a mouse, ensuring that navigation elements are properly labeled for screen reader users, and designing with sufficient color contrast for those with visual impairments.

- **Use of ARIA labels and landmarks**: Implementing **Accessible Rich Internet Applications** (**ARIA**) labels and landmarks can greatly enhance the accessibility of navigation. These tools provide users of assistive technologies with the context and information they need to navigate the site more effectively.

- **Consistent and predictable menus**: Consistency in the structure and behavior of navigation menus across the website can greatly aid both discoverability and accessibility. Users learn the navigation patterns of a site as they use it; maintaining these patterns consistently reduces cognitive load and makes the site more navigable.

- **Testing with diverse user groups**: Regularly testing the website's navigation with diverse user groups, including those with disabilities, is crucial. This testing can reveal hidden issues in both discoverability and accessibility and provide insights into how the navigation experience can be improved.

- **Responsive and adaptive navigation**: In a multi-device world, ensuring that navigation is responsive and adapts effectively to different screen sizes and input methods (such as touchscreens) is key to both discoverability and accessibility. Navigation should remain intuitive and accessible whether a user is on a desktop, tablet, or mobile device.

In conclusion, enhancing discoverability and accessibility in web navigation is about creating a user experience that is intuitive, inclusive, and effective. By focusing on clear navigation labeling, logical structure, adherence to accessibility standards, and regular user testing, websites can ensure that their content is reachable and usable by the widest possible audience. This approach not only improves user satisfaction but also aligns with best practices in web design and development.

As with anything in software development, we need to explore testing properly.

Testing and iteratively improving navigation

The importance of testing and iterative improvement in navigation cannot be overstated. Effective navigation is not just about initial design; it's an ongoing process of adaptation and refinement. Regular testing and iteration are essential in ensuring that the navigation remains intuitive, user-friendly, and aligned with evolving user needs and behaviors.

Regular testing is crucial for identifying issues in navigation that might not be immediately apparent during the design phase. This includes testing for usability, accessibility, and responsiveness across different devices and browsers. User testing can reveal how real users interact with the navigation, providing insights that analytics alone might not uncover.

Various methods can be employed for testing navigation. User testing sessions, where real users interact with the navigation and provide feedback, are invaluable. A/B testing can be used to compare different navigation structures and determine which is most effective. Heatmaps and analytics can also provide data on how users interact with the navigation, highlighting areas that might need improvement.

User feedback is a critical component of the iterative process. It provides direct insight into user experiences and preferences. Gathering feedback can be done through surveys, feedback forms, or during user testing sessions. Actively seeking and incorporating this feedback ensures that the navigation evolves in response to user needs.

Navigation design should be approached as an iterative process. Based on testing results and user feedback, changes and improvements should be made regularly. This process involves not only fixing issues but also experimenting with different approaches to find the most effective navigation solution.

As part of the iterative process, regularly testing for accessibility is crucial. This ensures that the navigation is usable by all individuals, including those with disabilities. Tools such as screen readers and accessibility testing software can aid in this process.

Navigation not only needs to be user-friendly but also fast and responsive. Performance testing is an important part of the iteration process, ensuring that the navigation elements load quickly and function smoothly across all devices and connections.

Continuously monitoring key performance metrics such as bounce rate, time on site, and conversion rates can provide ongoing insights into the effectiveness of the navigation. Changes in these metrics can indicate when further iterations or testing may be required.

The digital landscape is constantly evolving, and navigation should adapt accordingly. Staying informed about new trends, technologies, and user behaviors is essential for ensuring that the navigation remains current and effective.

In conclusion, testing and iterative improvement are fundamental in creating and maintaining effective web navigation. By regularly testing, incorporating user feedback, and staying attuned to changing trends, you can ensure that your website's navigation continues to meet and exceed user expectations, ultimately enhancing the overall user experience and effectiveness of your website's conversion pathways.

Summary

As we conclude *Chapter 7*, let's reflect on the essential insights and skills we've acquired. This chapter has been instrumental in deepening our understanding of effective web navigation and its pivotal role in user experience and conversion optimization.

We've explored the core principles of user-friendly navigation, learning how to organize and structure website navigation in a way that is intuitive and effortless for users. This skill is crucial in ensuring that visitors can navigate your site with ease, significantly improving the overall user experience.

A key focus of this chapter was on understanding and meeting user expectations around mobile navigation. With the increasing dominance of mobile browsing, mastering mobile navigation design is more important than ever. We've learned how to create navigation that is not only responsive but also caters to the unique needs and constraints of mobile users.

Additionally, we've examined how conversion paths differ across devices. This understanding is vital for designing user journeys that are optimized for engagement and conversions, regardless of the device being used. By tailoring these paths to different devices, we can maximize the effectiveness of our websites in driving desired actions.

Another important aspect we covered was identifying and rectifying common navigational design flaws. Addressing these flaws is essential for enhancing website usability and performance, reducing user frustration, and improving overall site effectiveness.

The skills and knowledge you've gained from this chapter are invaluable in creating websites that are not only visually appealing but also highly functional and user-centric. By focusing on intuitive navigation, responsive design, and effective conversion pathways, we can significantly enhance the user experience and drive better engagement and conversions.

In the next chapter, we will build upon these concepts by delving into the importance of making our websites accessible to all users, including those with disabilities. This is a natural progression from our focus on user-friendly navigation as accessibility is a key component of usability. We'll explore why web accessibility is not just a moral obligation but also a strategic business decision that can enhance user experience, broaden your audience, and improve overall site performance.

8

The Business Case for Web Accessibility

In this chapter, we'll embark on a comprehensive journey to uncover the multifaceted benefits that accessible web practices bring to businesses. Beyond mere compliance with legal standards, this chapter illuminates how adopting a proactive stance to accessibility can catalyze a broader spectrum of advantages, from expanding market reach to fostering a natural synergy with **responsive web design (RWD)** and realizing both tangible and intangible returns on investment.

This exploration is not just about adhering to regulations but recognizing accessibility as a cornerstone of modern web development that aligns closely with ethical business practices and competitive advantage. We'll delve into how inclusive designs not only welcome a wider audience but also enhance the **user experience (UX)** for all, creating a more engaging and usable web environment. Through a blend of theory, practical insights, and compelling case studies, you will gain a deep understanding of how web accessibility transcends compliance to become a key driver of innovation, customer loyalty, and business growth.

As we dissect the natural alignment between RWD and accessibility, we'll uncover how these disciplines, when integrated, offer a seamless web experience across a multitude of devices and user abilities, further emphasizing the importance of a user-centric design philosophy. Moreover, this chapter aims to equip you with the knowledge to recognize the direct and indirect ROI of implementing accessible web practices, from legal risk mitigation and enhanced SEO to increased user satisfaction and brand reputation.

By the end of this chapter, you will not only appreciate the ethical imperatives of web accessibility but will also be convinced of its strategic value, armed with the insights needed to champion accessibility within your organization. This chapter sets the stage for a deeper dive into how businesses can — and have — transformed the challenge of accessibility into opportunities for innovation and expanded market presence.

The following topics will be covered in this chapter:

- Accessibility beyond compliance
- Broadening market reach with inclusive design
- RWD and accessibility — a harmonious relationship
- ROI in accessible web practices

Accessibility beyond compliance

At its core, web accessibility is a matter of ethics. It's about recognizing the intrinsic right of all individuals, regardless of their physical or cognitive abilities, to participate fully in all aspects of society. This includes the digital realm, which has become intrinsically woven into the fabric of daily life. By integrating accessibility principles into web design and content, businesses can acknowledge and act upon their responsibility to eliminate barriers to digital information and tools.

Accessibility intersects with broader **corporate social responsibility** (CSR) initiatives, where businesses are increasingly held accountable for their impact on societal well-being. Implementing accessible web practices demonstrates a company's commitment to CSR by actively contributing to a more equitable society. It signals to stakeholders — from customers to employees to investors — that the company values diversity, equity, and inclusion.

The pursuit of accessibility is fundamentally about building an inclusive digital world. As the internet becomes the cornerstone for accessing education, employment, commerce, healthcare, and government services, ensuring these resources are universally accessible becomes paramount. Businesses that prioritize web accessibility are at the forefront of this movement, leading by example and setting benchmarks for others to follow.

Companies that champion web accessibility often find that their efforts have a ripple effect, inspiring suppliers, partners, and competitors to elevate their standards for inclusivity. This leadership fosters a competitive environment where the benchmarks for success include not just financial performance but also social impact. By advocating for accessibility, businesses can drive industry-wide changes, contributing to global efforts toward inclusivity and equality.

Beyond the immediate ethical considerations, prioritizing accessibility can enhance a company's long-term sustainability and brand perception. Consumers are increasingly aligning their loyalties with brands that demonstrate social responsibility and ethical practices. Therefore, accessibility becomes a key differentiator in the marketplace, enhancing brand loyalty among a diverse consumer base and improving employee morale by fostering a culture of inclusivity.

In conclusion, the ethical implications and social responsibility associated with web accessibility go far beyond mere legal compliance. They touch on the very essence of what it means to conduct business ethically in the digital age. By embracing accessibility, companies not only adhere to moral and social obligations but also contribute to a more inclusive and equitable society. This commitment to accessibility, rooted in ethical considerations, strengthens the business case for web accessibility, showcasing it as an integral component of responsible business practice.

Now, let's explore how we can make better experiences for every user.

Enhancing the UX for all

A pivotal point is the universal benefit of enhancing the UX for all through accessibility. This section delves into how inclusive design practices not only meet the needs of individuals with disabilities but also improve the digital experience for a broader audience, underscoring the intrinsic value of accessibility beyond the realm of compliance.

The concept of universal design is at the heart of creating an enhanced UX for all users. By adhering to principles that advocate for the broad usability of environments, products, and services, businesses can craft web experiences that are inherently more accessible. This approach ensures that websites are navigable, understandable, and usable for everyone, regardless of their abilities or the devices they use to access the internet.

Accessibility initiatives often lead to the simplification of web interfaces, which benefits all users. Complex navigations, cluttered interfaces, and convoluted processes can hinder the UX significantly. By streamlining design elements and interactions to accommodate users with disabilities, websites inadvertently become more user-friendly for everyone. Clear headings, consistent navigation structures, and straightforward tasks contribute to a smoother, more intuitive web experience.

The overlap between accessible web practices and improved **search engine optimization (SEO)** enhancements is considerable. Many accessibility measures, such as providing alternative text for images and ensuring content is organized logically through headings, also improve a site's SEO. This makes content more discoverable to users through search engines, amplifying the reach of the website to a wider audience and driving increased engagement.

Optimizations for accessibility, such as efficient coding and reduced reliance on heavy multimedia elements, can decrease web page load times. Faster load times enhance the UX for all users, particularly those accessing content on mobile devices or in areas with limited internet connectivity. This responsiveness is crucial for keeping users engaged and reducing bounce rates.

Accessibility practices give users more control over how they interact with digital content. Features such as text resizing, high-contrast modes, and keyboard navigation options allow users to customize their browsing experience to suit their preferences and needs. This level of flexibility is beneficial not only for users with specific accessibility requirements but for all users who want a more personalized web experience.

Inclusive design emphasizes clear feedback and error prevention, which are critical for users with cognitive disabilities. This focus results in interfaces where mistakes are minimized, and users are guided more clearly through processes such as form filling and transactions. Such enhancements in feedback and error handling improve the UX for all users by making digital interactions more forgiving and less frustrating.

The enhancement of UX for all through accessibility is a testament to the inclusive potential of digital spaces. When businesses commit to web accessibility, they invest in a philosophy that values every user's experience. This investment transcends compliance, reaching into the essence of what it means to create truly user-centric digital products and services. In doing so, businesses not only advocate for a more inclusive internet but also realize the extensive benefits of a universally positive UX.

Outside of accessibility being a driver for inclusivity, it's also a driver for innovation.

Accessibility as a driver for innovation

In the landscape of digital development, accessibility is often seen through the lens of compliance, a box to be checked. However, when we delve deeper, we uncover a compelling narrative: accessibility as a driver for innovation. This perspective shifts the paradigm, positioning accessibility not as a mere obligation but as a catalyst for creative solutions and technological advancements.

Accessibility challenges designers and developers to think outside the conventional boundaries of web design. The necessity to create digital experiences that can be used by people with a wide range of abilities prompts innovation. It encourages the exploration of new technologies, frameworks, and design philosophies. For instance, voice navigation and gesture controls, initially developed to accommodate users with disabilities, have found broader applications, enhancing the UX for a wide audience and opening new avenues for interaction with digital content.

Investing in accessibility drives research and development efforts toward inclusive technologies. It pushes companies to allocate resources to discover and implement solutions that address the diverse needs of users. This R&D commitment not only leads to breakthroughs in accessible design but also propels the development of products and services that stand out in the market for their usability and user-centric approach.

Incorporating accessibility into the Agile development process exemplifies its role as an innovation driver. By integrating accessibility considerations from the outset, teams are encouraged to adopt flexible, iterative approaches to problem-solving. This integration ensures that accessibility is not an afterthought but a key consideration that influences the development life cycle, leading to more innovative and inclusive products.

Designing for accessibility broadens the scope of user engagement by making digital experiences more adaptable to individual user needs. This adaptability not only caters to users with disabilities but also meets the varying preferences and contexts of all users. By doing so, accessibility-driven innovation enhances the capacity of digital products to engage a wider audience, making these products more versatile and valuable.

There are various real-world examples where accessibility considerations have led to innovative products and features. Social media platforms introducing alt text for images, streaming and video platforms implementing audio descriptions or auto-generated captions, and websites offering customizable display settings are just a few instances where accessibility initiatives have enhanced the overall UX and set new industry standards.

Viewing accessibility as an innovation driver shifts the narrative from compliance to competitive advantage. It challenges the status quo, inspiring businesses to explore creative solutions that not only make the web more inclusive but also push the boundaries of what is possible in digital design and development. This approach not only fulfills a social responsibility but also positions companies as leaders in a tech-driven world where inclusivity becomes a benchmark of excellence.

Positioning your company as a leader helps create returning customers who trust your brand.

Building brand reputation and loyalty

The impact of accessibility on building brand reputation and loyalty is profound and multifaceted. Far from being a mere compliance checklist, accessible web practices offer a significant opportunity to enhance a brand's standing in the marketplace and foster deep, lasting connections with consumers.

Commitment to web accessibility reflects a brand's dedication to inclusivity and equality, values that are increasingly important to consumers. By ensuring that digital services and content are accessible to all, including those with disabilities, companies send a powerful message about their corporate ethics and social responsibility. This commitment can elevate brand perception, distinguishing the company as a socially conscious leader in its field.

Accessible web design opens the door to a wider audience, including the millions of people worldwide who have disabilities. It also caters to the aging population and those with temporary impairments. By removing barriers to access, companies can significantly expand their customer base. This inclusivity not only boosts market reach but also enhances customer diversity, which can lead to richer customer insights and innovation.

Accessibility improvements often lead to a better UX for all customers. Features such as clear navigation, legible fonts, and alternative text for images improve the overall usability of a website. A positive UX is a key driver of customer satisfaction, which, in turn, fosters brand loyalty. When customers have a seamless interaction with a brand online, they are more likely to return and recommend the brand to others.

Consumers are increasingly aligning their spending with their values, showing loyalty to brands that demonstrate a commitment to social causes, including accessibility. By prioritizing web accessibility, companies can strengthen their relationships with socially conscious consumers. This alignment of brand values with customer values is a powerful driver of loyalty, encouraging customers to form long-term relationships with the brand.

Conversely, failure to address web accessibility can result in negative publicity, legal challenges, and a damaged brand reputation. Proactively embracing accessibility demonstrates a brand's commitment to doing the right thing, which can protect against the reputational damage associated with accessibility lawsuits and complaints.

Many brands have strengthened their market position through a focus on accessibility. These companies often report not just an increase in customer satisfaction among users with disabilities but an overall uplift in brand perception and customer loyalty across their entire customer base. By highlighting these success stories, we can see the tangible benefits that accessibility brings to brand reputation and loyalty.

In conclusion, strategically integrating accessibility into web practices offers a powerful avenue for building brand reputation and loyalty. It's a clear win-win: businesses enhance their market reach and brand perception, while consumers benefit from a more inclusive and user-friendly digital environment. As such, web accessibility should be seen as an essential component of a brand's identity and a key factor in its long-term success.

Let's explore a case study.

Case studies — success stories of accessible design

Exploring case studies and success stories of accessible design not only highlights the tangible benefits of such initiatives but also serves as a beacon for companies who are contemplating integrating accessibility into their digital strategy. These narratives demonstrate the profound impact that accessible design can have on user engagement, brand loyalty, and ultimately, business success.

So, let's begin:

- **Global technology leader — Microsoft**

  Microsoft's commitment to accessibility is evident across its product range, from Windows to Office and beyond. The company's inclusive design principle has led to innovations such as the Narrator screen reader, which provides a voiceover for users with visual impairments, and the Xbox Adaptive Controller, designed for gamers with limited mobility. These efforts have not only opened up new markets for Microsoft but also solidified its reputation as a leader in tech innovation and inclusivity. The brand loyalty among users who benefit from these features is strong, with many considering Microsoft products the only viable option for their needs.

- **Retail giant — Target**

  After facing a lawsuit in 2006 for its inaccessible website, Target undertook a comprehensive overhaul of its digital properties to ensure they were accessible to all users, including those with visual impairments. This transformation included implementing screen-reader-friendly designs, keyboard navigation, and alternative text for images. This revamp not only mitigated legal risks but also resulted in a surge in customer satisfaction scores and online sales, showcasing the direct ROI of accessible design.

- **Banking sector — Barclays**

  Barclays has set a gold standard in the finance industry for accessibility, viewing it as a key aspect of customer service. Their digital platforms are designed with accessibility in mind, offering features such as high-contrast modes and text resizing to accommodate users with visual impairments. Barclays' commitment to accessibility extends beyond its digital presence, with initiatives such as talking ATMs and braille statements. This dedication has earned Barclays numerous awards and, more importantly, the loyalty of a diverse customer base that values inclusivity.

- **E-commerce — Etsy**

  Etsy, the global online marketplace for handmade goods and vintage items, has made significant strides in web accessibility, recognizing its importance in fostering a diverse community of buyers and sellers. By focusing on key areas such as color contrast, keyboard navigation, and descriptive link text, Etsy has improved its platform's usability for people with disabilities. These improvements have not only enhanced the shopping experience for all users but also increased the diversity of the marketplace, contributing to Etsy's growth and brand strength.

- **Public sector — GOV.UK**

  The UK government's official website, GOV.UK stands as a paragon of accessible design in the public sector. By adhering strictly to **Web Content Accessibility Guidelines** (**WCAG**) and focusing on simplicity and clarity, GOV.UK ensures that all citizens, regardless of ability, can access government information and services online. This approach has vastly improved public satisfaction with government digital services and set a benchmark for accessible design in the public domain.

Each of these case studies underscores a crucial message: that accessible design is not just about meeting compliance standards but about unlocking innovation potential for reaching wider audiences and building a more inclusive digital world. These success stories highlight the positive outcomes of accessible design, from enhanced user engagement and satisfaction to increased sales and brand loyalty, offering compelling evidence of the business case for web accessibility.

Now, it's time to broaden our market research.

Broadening market reach with inclusive design

Understanding inclusive design is pivotal for businesses aiming to broaden their market reach. Inclusive design is a methodology that considers the full range of human diversity to make products, services, and environments accessible to as many people as possible. This approach not only addresses the needs of individuals with disabilities but also caters to older populations, those with temporary injuries, and others who might face barriers in using digital platforms.

At its core, inclusive design acknowledges that people experience and interact with the world in diverse ways. It challenges designers to think beyond the average user and create solutions that cater to previously unconsidered needs. This means developing websites and digital tools that can be used by people with a wide array of hearing, movement, sight, and cognitive abilities.

Inclusive design can be viewed through three primary dimensions:

- **Recognizing diversity and uniqueness**: Every user brings a unique set of experiences, preferences, and needs to the table. Inclusive design respects these differences and seeks to accommodate them rather than forcing conformity to a narrow set of criteria.

- **Inclusive process and tools**: The process of design itself should be inclusive. This involves engaging with a range of users from the outset and incorporating their feedback directly into the design process. Tools and methodologies that are used in design and development should also support inclusivity, allowing for flexibility and adaptation.

- **Broader beneficial impact**: Solutions that are created under the principles of inclusive design often have a broader beneficial impact, offering improvements for a wide audience beyond those with specific accessibility needs. For example, subtitles on videos benefit not only individuals who are deaf or hard of hearing but also those watching in noisy environments or who are non-native language speakers.

Implementing inclusive design requires a shift in perspective. It involves the following aspects:

- **Empathy and awareness**: Understanding the challenges faced by diverse users and developing empathy for their experiences.

- **Broad user engagement**: Actively engaging with a wide range of users during the design and testing phases to gather insights and feedback.

- **Flexibility and adaptability**: Designing systems that are flexible and adaptable, allowing users to customize their experiences to meet their needs.

- **Ongoing learning and adaptation**: Recognizing that inclusivity is a moving target as technology and societal norms evolve. Continuous learning and adaptation are required to meet changing needs.

Understanding and implementing inclusive design is a strategic move that extends the reach of digital platforms to encompass a wider, more diverse audience. It not only fulfills ethical obligations toward accessibility but also opens up new markets and opportunities for innovation. By embracing the principles of inclusive design, businesses can create more usable, desirable, and accessible products and services, thereby broadening their market reach and enhancing their competitive edge in an increasingly diverse global market.

When thinking about accessibility, it's important to think about the users who have disabilities and how you can help them.

Demographics of disability

Recognizing the sheer diversity and scale of the audience that benefits from accessible web practices reveals the vast, often untapped market potential that inclusivity unlocks.

Unpacking the demographics

The demographics of disability encompass a wide spectrum of the global population. According to the **World Health Organization (WHO)**, over a billion people, approximately 15% of the world's population, experience some form of disability. This number is not static; it grows due to factors such as aging populations, the prevalence of chronic health conditions, and advancements in medical care that increase survival but may leave individuals with lasting impairments.

Aging populations

A significant driver of disability demographics is the aging population. Older adults often experience a range of age-related impairments, including reduced mobility, vision and hearing loss, and cognitive changes. So, designing accessible digital experiences becomes crucial in ensuring that the growing number of older users remains connected and engaged with digital content and services.

Temporary and situational disabilities

Beyond the static numbers are the dynamic conditions of temporary and situational disabilities. Temporary disabilities, such as a broken arm, and situational limitations, such as bright sunlight glare on a screen, affect how individuals interact with digital platforms. Inclusive design accounts for these varied experiences, ensuring that web accessibility addresses not just permanent but also temporary and situational challenges.

Economic impact and market reach

The economic impact of ignoring the needs of people with disabilities is substantial. By not considering accessible design, businesses risk alienating a significant portion of their potential market. Conversely, companies that embrace accessibility can tap into a market segment with considerable spending power. In the US alone, the disposable income of adults with disabilities is estimated to be in the billions, underscoring the economic incentive behind inclusive design practices.

The ripple effect of inclusive design

Inclusive design resonates beyond individuals with disabilities, benefiting a broader audience. Features that make websites accessible, such as clear navigation, legible fonts, and alternative text for images, improve the overall UX for everyone. This universality of accessible design underscores its value in reaching diverse user bases and enhancing usability for all demographics.

The demographics of disability illustrate a diverse and substantial audience that demands recognition in the digital arena. Understanding the scale and diversity of this audience is not just about acknowledging their rights to access information and services; it's about recognizing their value as consumers, users, and participants in the digital world. Therefore, inclusive design is not merely a compliance or ethical issue but a strategic business decision that broadens market reach and opens up new avenues for engagement and growth. By prioritizing accessibility, businesses can ensure they cater to the full spectrum of human diversity, unlocking the full potential of the global market.

Case studies — companies leading with inclusive design

It is imperative to spotlight certain case studies. These narratives not only serve as compelling testimonials to the efficacy and impact of inclusive practices but also illustrate how businesses can thrive by embracing a philosophy that caters to all users. Here, we'll delve into a selection of companies whose commitment to inclusive design has not only broadened their market reach but has also positioned them as pioneers of accessibility in the digital age:

- **Airbnb — cultivating inclusivity in global travel**

  Airbnb's commitment to inclusivity has transformed the travel industry, making it more accessible to people with disabilities. The platform's "Accessible Travel" features allow users to filter search results based on a range of accessibility needs, including step-free access and entryways wide enough for wheelchairs. By prioritizing accessibility in its listings, Airbnb has not only expanded its market reach but also fostered a more inclusive global travel community.

- **LEGO — the building blocks of inclusive play**

  LEGO's Braille Bricks initiative is a testament to the power of inclusive design in transforming traditional play. Designed to help children with vision impairments learn braille playfully and engagingly, Braille Bricks underscore LEGO's commitment to ensuring that play is accessible to every child. This innovative approach has not only garnered global acclaim but has also broadened LEGO's market reach, reinforcing the brand's position as a leader in inclusive play.

- **Spotify — harmonizing music with accessibility**

  Spotify has set a high standard for accessibility in digital music services. The platform offers features such as high-contrast themes, screen reader compatibility, and keyboard navigation, making it accessible to users with visual impairments. By integrating accessibility into its design ethos, Spotify has ensured that everyone can experience the joy and connection of music, thereby expanding its user base and strengthening brand loyalty among a diverse audience.

These case studies exemplify the profound impact that inclusive design can have on a company's market reach and brand reputation. Airbnb, LEGO, and Spotify demonstrate that accessibility is not just a moral imperative but a strategic business advantage. Their success stories provide a roadmap for other companies aspiring to make inclusivity a cornerstone of their business model. In embracing inclusive design, businesses can unlock untapped market potential, foster brand loyalty, and contribute to a more inclusive world.

Inclusive design and digital innovation

The correlation between inclusive design and digital innovation emerges as a compelling chapter that underscores the transformative power of inclusivity in driving technological advancement. This section elucidates how adopting inclusive design principles not only caters to a broader audience but also serves as a catalyst for innovation, pushing the boundaries of what's possible in the digital realm.

Let's explore the importance of inclusive design.

The catalyst of inclusion

Inclusive design challenges the status quo by requiring creators to think beyond traditional user archetypes and consider the full spectrum of human diversity. This approach necessitates a deep understanding of various user needs and preferences, which, in turn, fosters a culture of innovation. Companies that embrace this challenge often discover novel solutions that benefit all users, not just those with disabilities.

Broadening perspectives for broader solutions

Inclusion acts as a lens that broadens the perspective of designers and developers, encouraging them to explore a wider array of solutions. For instance, voice-controlled assistants, initially designed to aid users with mobility and visual impairments, have become mainstream, simplifying digital interactions for a wide user base. Similarly, captioning and sign language in video content, aimed at deaf and hard-of-hearing audiences, enhance comprehension for non-native language speakers and environments where audio isn't feasible.

Technological breakthroughs stemming from accessibility needs

Many technological breakthroughs have their roots in addressing accessibility challenges. Screen readers, text-to-speech technology, and tactile feedback devices were developed to make digital content accessible to users with visual impairments. These technologies have broader applications, enhancing digital experiences for all users and driving innovation in user interface design.

Inclusive design as a competitive edge

In the digital marketplace, innovation serves as a key differentiator. Companies that integrate inclusive design into their product development processes often achieve a competitive edge. They not only meet the needs of a wider demographic but also anticipate changes in user behavior and regulatory environments. This proactive approach to design positions companies as leaders in UX, capable of capturing emerging markets and responding dynamically to user needs.

Collaboration and co-creation

Inclusive design encourages collaboration between stakeholders with diverse perspectives, including users with disabilities, designers, developers, and industry experts. This collaborative environment fosters co-creation, where solutions are developed not just for but with users. Such an approach leads to richer, more effective digital innovations that are finely tuned to user needs.

Driving sustainability in digital products

Innovations born from inclusive design often emphasize sustainability as products that have been created to be accessible and usable for everyone tend to have longer life cycles and broader applicability. This sustainability reduces the need for frequent redesigns or updates, contributing to more stable and enduring digital products.

Inclusive design and digital innovation are inextricably linked, with inclusivity acting as a springboard for creative solutions that redefine UXs. By embedding inclusive practices at the heart of digital product development, companies can unlock new levels of innovation, fostering technologies that not only accommodate but celebrate human diversity. This commitment not only broadens market reach but also propels the digital world toward a more inclusive future, where technology serves everyone equitably.

Strategies for implementing inclusive design

Implementing inclusive design is a strategic process that requires thoughtful planning and execution. It's about creating digital experiences that are accessible and meaningful to as many people as possible, regardless of their abilities or circumstances. To effectively broaden market reach with inclusive design, businesses must adopt a comprehensive approach that integrates inclusivity into every stage of the design and development process.

Here are some key strategies for implementing inclusive design:

- **Start with awareness and training**

  The first step in implementing inclusive design is building awareness within the organization about what inclusive design is and why it matters. Training sessions for design, development, and content teams can instill a deep understanding of inclusive design principles and practices. This foundational knowledge ensures that everyone involved in product creation is equipped to consider accessibility from the outset.

- **Involve diverse users early and often**

  Genuine inclusivity comes from engaging with a diverse group of users, including people with disabilities, throughout the design process. This can be achieved through user research, interviews, and testing sessions that involve participants with a range of abilities. Feedback from these users is invaluable in identifying potential barriers and ensuring that digital products are designed with real-world needs in mind.

- **Adopt a universal design approach**

  Universal design is the practice of creating products and environments that can be used by all people, to the greatest extent possible, without the need for adaptation or specialized design. By adopting universal design principles, companies can create more accessible products from the start. This approach not only makes products more inclusive but also often leads to innovations that improve usability for all users.

- **Prioritize accessibility in the development life cycle**

  Incorporating accessibility considerations into the development life cycle requires making it a priority from the initial concept through to launch and beyond. This includes setting clear accessibility goals, using accessibility guidelines (such as WCAG) as benchmarks, and integrating accessibility checks into regular testing procedures. Tools and automated tests can identify some issues, but manual testing by users with disabilities is also crucial.

- **Embrace flexibility and customization**

  An effective inclusive design strategy allows for flexibility and customization, enabling users to adjust their experience according to their preferences and needs. Features such as adjustable text sizes, color contrast options, and alternative navigation methods can make digital products more accessible to a broader audience.

- **Foster a culture of continuous improvement**

  Inclusive design is an ongoing process. Technologies and user needs evolve, and what works well today may need adjustment in the future. Creating a culture of continuous improvement, where feedback is actively sought and used to make iterative enhancements, is key to maintaining and improving accessibility over time.

- **Measure impact and iterate**

  To understand the effectiveness of inclusive design efforts, it's important to measure their impact. This can include analyzing user engagement metrics, conducting usability studies with diverse users, and gathering qualitative feedback. Insights from these activities should inform future design iterations, ensuring that digital products continue to meet the needs of a wide range of users.

Implementing inclusive design is a strategic endeavor that can significantly enhance a company's market reach. By embedding accessibility and inclusivity into the fabric of the design and development process, businesses can create digital experiences that are not only compliant with legal standards but also genuinely welcoming to everyone. This approach not only opens up new markets but also positions companies as leaders in innovation and social responsibility.

Next, we'll explore the direct link between RWD and accessibility.

RWD and accessibility — a harmonious relationship

In exploring the symbiotic relationship between RWD and web accessibility, it's crucial to begin by laying down the foundations of RWD. RWD is a web development approach that creates dynamic changes to the appearance of a website, depending on the screen size and orientation of the device being used to view it. This approach ensures that a site is accessible and usable on a wide range of devices, from desktop monitors to mobile phones, thereby enhancing UX and accessibility.

Core principles of RWD

To recap, here are some of the core principles of RWD:

- **Fluid grids**: One of the cornerstones of RWD is the use of fluid grid systems. Unlike traditional layouts, which use fixed-width measurements, fluid grids use relative units such as percentages, which allow the layout to adapt to the user's viewing environment. This flexibility is fundamental to creating web pages that adjust seamlessly across different screen sizes.

- **Flexible images and media**: Just as fluid grids allow you to dynamically resize layout components, flexible images and media ensure that these elements also respond to changes in screen size. This adaptability prevents images and videos from exceeding their containing elements, which can disrupt the layout and the user's ability to interact with the page effectively.

- **Media queries**: Media queries are a critical feature of CSS that enables web designers to apply styles based on the specific characteristics of the device viewing the site, such as its width, resolution, and orientation. By using media queries, designers can create distinct layouts for different devices, ensuring that content is presented in the most optimal format possible.

Now, let's see how we can improve accessibility through RWD.

Enhancing accessibility through RWD

The foundational elements of RWD naturally align with the goals of web accessibility. By designing websites to be flexible and adaptable, RWD addresses several accessibility challenges by default:

- **Visibility**: Fluid grids and flexible media ensure that content can be easily viewed on screens of varying sizes, which is particularly beneficial for users with low vision who may rely on screen magnification tools

- **Control**: Media queries allow for the creation of device-specific interactions, improving the usability of websites for people using a range of input methods, including touchscreens and keyboards

- **Consistency**: RWD maintains consistency in navigation and interface elements across devices, aiding users with cognitive disabilities in understanding and navigating content

- **Customization**: The adaptability of responsive design supports personalization needs, such as adjusting text sizes and colors, which can be crucial for users with visual impairments or reading disabilities

The foundations of RWD are not merely technical solutions to the challenges of multi-device web browsing; they are also intrinsic to creating accessible digital spaces. RWD and web accessibility go hand in hand, each enhancing the efficacy of the other. By understanding and implementing the core principles of RWD, web designers, and developers can create websites that are not only universally usable across devices but also inherently more accessible to people with a wide range of abilities. This harmonious relationship between RWD and accessibility underscores the importance of adopting a holistic approach to web design where flexibility, adaptability, and user-centricity are paramount.

Now that we understand how to enhance accessibility, let's explore the intersection between the two.

The intersection of rWD and web accessibility

The intersection of RWD and web accessibility is critical as it's where the principles of creating adaptable and flexible digital environments converge to ensure these environments are accessible to all users, including those with disabilities. This intersection is not merely about technical compliance but fostering a deeper understanding of how RWD can enhance accessibility and, by extension, the overall UX.

Let's break down this intersection and look at it in more detail.

Complementary goals

RWD and web accessibility share complementary goals: to provide users with the best possible experience, regardless of their device or abilities. RWD achieves this by ensuring that websites look and function well on any screen size, while web accessibility focuses on making content usable for people with a wide range of abilities. When combined, these approaches ensure that a website is not only flexible and adaptive but also inclusive.

Enhancing usability for all

The synergy between RWD and web accessibility inherently enhances usability. For instance, RWD's approach to flexible grids and images ensures that content is legible and navigable on small screens without the need for horizontal scrolling, which is also a boon for users with mobility impairments who may find scrolling difficult. Similarly, the use of media queries to adjust layout and content presentation can be leveraged to introduce accessibility features for different user preferences, such as high-contrast modes for users with visual impairments.

Streamlined navigation

RWD's emphasis on streamlined and consistent navigation across devices aligns with accessibility guidelines that call for intuitive, straightforward navigation mechanisms. By designing menus and navigation structures that are easy to use on both desktop and mobile devices, designers can improve the site's usability for everyone, including people who use screen readers or rely solely on keyboard navigation.

Unified design approach

At the heart of RWD and web accessibility is a unified design approach that considers the UX holistically. This approach entails creating content and designs that are flexible and adaptable, not as an afterthought or for a specific group of users, but as a fundamental aspect of the web design process. This means thinking about how content reflows on different screen sizes, how color schemes affect readability, and how interactive elements function across input methods.

Overcoming challenges with innovative solutions

The intersection of RWD and web accessibility encourages innovation in tackling design and usability challenges. For example, the need to make interactive elements such as drop-down menus accessible both on touchscreens and for keyboard users can lead to the development of new, more intuitive interface components. This drive for innovation not only solves accessibility challenges but often results in a better product for all users.

The intersection of RWD and web accessibility is fertile ground for innovation and improvement in digital experiences. By embracing the principles of both disciplines, web professionals can create digital environments that are not only versatile and responsive to the vast array of devices in use today but also accessible and inclusive for users of all abilities. This convergence not only amplifies the usability and reach of web content but also underscores the ethical and practical importance of building a web that truly is for everyone. The harmonious relationship between RWD and web accessibility exemplifies how designing with empathy and consideration for all users leads to richer, more engaging digital experiences.

Challenges and opportunities in combining RWD with accessibility

The fusion of RWD with web accessibility presents a unique set of challenges and opportunities for web professionals. Let's take a closer look.

Challenges in integrating RWD and accessibility

First, we will look at the challenges in integrating RWD and accessibility. Let's begin:

- **Maintaining accessibility across devices**: One of the most significant challenges is ensuring consistent accessibility across various devices with differing screen sizes, resolutions, and interaction models. While RWD focuses on visual and functional adaptation, integrating accessibility requires a nuanced approach to ensure features such as keyboard navigability and screen reader compatibility are maintained uniformly.

- **Complex navigation structures**: RWD often simplifies complex navigation to fit smaller screens, which can inadvertently obscure content or make navigation more challenging for users with disabilities. Ensuring that these streamlined navigation menus remain accessible and intuitive across all devices demands careful planning and design.

- **Dynamic content and interaction**: Responsive sites frequently employ dynamic content and interactions that adjust based on the user's device. Ensuring these dynamic elements are accessible, such as providing appropriate **Accessible Rich Internet Applications (ARIA)** roles and ensuring dynamic content updates are announced by screen readers, adds another layer of complexity to the design process.

- **Performance optimization**: Balancing performance with accessibility in RWD can be challenging. High-resolution images and multimedia, which are often scaled down for mobile devices, need alternative text descriptions and must be optimized to not hinder load times, both of which are crucial for users with slow connections and also impact overall accessibility.

Next, we will look at the various opportunities.

Opportunities stemming from convergence

Let's explore some of the opportunities that are presented as a result of convergence. These opportunities need to be identified and acted upon to truly help maximize the benefits of convergence:

- **Universal design principles**: The challenges of combining RWD with accessibility underscore the value of universal design principles, which aim to create solutions that can be used by the widest range of people. This approach not only addresses specific user needs but also improves usability for a broader audience, driving innovation in user interface design.

- **Enhanced UX**: The focus on accessibility within RWD leads to cleaner, more intuitive interfaces that benefit all users. Features such as larger clickable areas, clear font choices, and thoughtful layout adjustments contribute to a more enjoyable and efficient UX, regardless of device or ability.

- **SEO benefits**: Accessible, responsive websites tend to perform better in search engine rankings. Search engines favor sites that provide a good UX, characterized by fast load times, mobile-friendliness, and content accessibility. Thus, the efforts to combine RWD with accessibility can also enhance a site's visibility and reach.

- **Legal compliance and market expansion**: Adhering to web accessibility guidelines not only mitigates legal risks but also opens up new markets. By catering to the needs of users with disabilities, organizations can tap into a wider customer base, demonstrating CSR and expanding their market reach.

- **Innovation through inclusion**: The necessity to address both responsiveness and accessibility encourages innovation, leading to the development of new technologies, frameworks, and design methodologies. This culture of inclusive thinking drives the web forward, making it a richer, more diverse space for all users.

The journey to harmonize RWD with web accessibility is fraught with challenges but is also ripe with opportunities for growth and innovation. By embracing these complexities, designers, and developers can create digital experiences that are not only more inclusive and accessible but also more engaging and effective for everyone. This endeavor not only meets the ethical mandate to make the web accessible to all but also leverages inclusivity as a driver for technological advancement and broader market engagement.

Best practices for accessible responsive design

Integrating accessible responsive design into digital platforms isn't just about adhering to standards; it's about crafting experiences that everyone can navigate and enjoy, regardless of their device or abilities. Understanding and applying best practices for accessible RWD is crucial for this chapter. These guidelines not only enhance the usability and accessibility of websites but also ensure that all users benefit from thoughtful, inclusive design choices.

Here are the aforementioned best practices:

- **Embrace a mobile-first approach**

  Starting the design process with mobile considerations prioritizes simplicity and performance, key aspects of accessibility. A mobile-first approach ensures that content is essential, navigation is streamlined, and page elements are optimized for touch interactions. This foundation makes it easier to scale designs to larger screens while maintaining accessibility features.

- **Use semantic HTML**

  Semantic HTML5 elements (`<nav>`, `<header>`, `<footer>`, `<article>`, and `<section>`) provide inherent meaning and structure to web content, making it more navigable and understandable for users and assistive technologies alike. Using these elements correctly is fundamental in conveying the structure and purpose of content across all devices.

- **Implement flexible layouts and text**

  Employ fluid grid layouts that use percentages rather than fixed units, allowing content to adapt smoothly to any screen size. Similarly, ensure text can resize without loss of content or functionality, accommodating users who need larger fonts. CSS media queries should be used to adjust layouts and text sizes based on the viewport's dimensions.

- **Ensure interactive elements are accessible**

  Buttons, links, and form controls should be easily identifiable and usable across all devices. This means larger touch targets on mobile devices to accommodate finger navigation and clear focus styles for keyboard users. All interactive elements should be operable through keyboard and touch, without requiring precise mouse control.

- **Provide alternative text for images**

  All images and non-text content should have descriptive alternative text, allowing screen reader users to understand their purpose in the context of the page. This practice benefits users on slow connections or those who disable images to save data as they can still understand the content without seeing the images.

- **Utilize ARIA landmarks and roles**

  ARIA landmarks and roles offer a way to communicate the structure and purpose of web content to assistive technologies. Use ARIA landmarks to identify regions of the page (for example, navigation, main content, and search), and roles to describe the function of interactive elements, enhancing the usability of web applications, especially in dynamic content situations.

- **Optimize multimedia for accessibility**

  Ensure that all video and audio content is accessible by providing captions, transcripts, and audio descriptions. This not only aids users who are deaf or have hearing impairments but also benefits those in environments where audio cannot be played. Responsive video players should adjust to fit the screen size without obscuring controls or content.

- **Test across devices and assistive technologies**

  Regular testing on a variety of devices and with different assistive technologies is vital to identify and address accessibility issues. This includes screen readers, magnification software, and speech recognition tools. User testing with individuals who have disabilities can provide invaluable insights into the real-world accessibility of a site.

- **Foster continuous learning and improvement**

  The landscape of web technology and accessibility standards is ever-evolving. Staying informed about new developments, tools, and techniques is crucial for maintaining and improving the accessibility of responsive designs over time.

By following these best practices, designers and developers can ensure that their responsive websites are not just visually and functionally flexible across devices but are also accessible to all users, including those with disabilities. Accessible responsive design transcends the traditional boundaries of web development, offering a more inclusive and equitable internet — one that acknowledges and accommodates the full spectrum of human diversity.

Case studies — success stories of accessible RWD

Examining case studies of success stories in accessible RWD illuminates the profound impact that thoughtful, inclusive digital strategies can have on users and businesses alike. These examples serve not only as proof of concept but also as inspiration, showcasing how the principles of accessibility and responsiveness can be seamlessly integrated to create experiences that are universally usable and enjoyable. Let's take a look:

- **BBC News**

  The BBC News website is a sterling example of accessible RWD in action. Recognizing the diverse needs of its global audience, the BBC implemented a responsive design that adjusts content layout based on the device used, ensuring readability and navigability across platforms. What sets the BBC apart is its commitment to accessibility: the site features keyboard navigation, screen reader compatibility, and options for users to change the contrast and text size. This approach has not only improved the experience for users with disabilities but has also made the site more usable for everyone, leading to increased engagement and user satisfaction.

- **GOV.UK**

  The GOV.UK website, the official portal for the United Kingdom government, has been lauded for its accessible, user-centric design. Focused on delivering government services and information with clarity, GOV.UK employs RWD to ensure its resources are available to all citizens, regardless of the device they use. The site's design prioritizes simplicity and functionality, with a strong emphasis on accessible features such as high-contrast modes, text resizing, and clear, straightforward navigation. GOV.UK's success lies in its ability to serve a wide demographic, demonstrating how government websites can lead by example in accessibility and responsive design.

- **Shopify**

  Shopify, a leading e-commerce platform, empowers online retailers to create shops that are both accessible and responsive. Through its default themes, which adhere to WCAG 2.0 standards, Shopify provides merchants with the foundation to build online stores that are navigable and usable by people with disabilities. The platform's focus on RWD ensures that these stores work seamlessly across devices, offering a consistent shopping experience that has contributed to the platform's widespread adoption and the success of its merchants.

- **Starbucks**

  Starbucks' commitment to digital accessibility and inclusive design is evident in its responsive website and mobile app, which allow customers to order coffee and find stores with ease. The company undertook significant efforts to ensure its digital offerings are accessible, including incorporating screen reader support, sufficient color contrast, and accessible online forms. These improvements were driven by a desire to serve all customers equitably, leading to increased customer loyalty and recognition of Starbucks as a leader in corporate accessibility initiatives.

These case studies of accessible RWD success stories highlight the tangible benefits of integrating inclusive design principles into digital products. By prioritizing accessibility and responsiveness, organizations such as the BBC, GOV.UK, Shopify, and Starbucks have not only expanded their market reach but have also fostered a more inclusive digital world. Their achievements underscore the importance of designing with all users in mind, proving that accessibility and responsiveness are not just ethical imperatives but also strategic business advantages that drive growth, engagement, and customer satisfaction.

Now, let's explore the potential ROI that accessibility could produce.

ROI in accessible web practices

Understanding the relationship between the costs and benefits of accessible web practices is crucial for organizations aiming to justify their investment in accessibility. This section aims to dispel common misconceptions and highlight the tangible returns that accessible design can bring to businesses.

The cost of implementing web accessibility can vary widely depending on several factors, including the complexity of the website, the current level of accessibility, and whether accessibility is considered from the outset of design and development or retrofitted into an existing site. Initial costs may involve the following:

- **Auditing and assessment**: Conducting thorough accessibility audits to identify current barriers

- **Training and education**: Investing in training for design, development, and content teams

- **Redesigning and development**: Making necessary adjustments to design and code to meet accessibility standards, such as WCAG

While these initial investments might seem daunting, they are typically offset by the broad range of benefits that accessible web practices bring.

These benefits include the following:

- **Enhanced UX and reach**: By making websites accessible, businesses can significantly widen their audience, reaching millions of users with disabilities. This expansion not only aligns with ethical standards but also taps into a market segment with considerable spending power.

- **Improved SEO performance**: Many accessibility best practices, such as providing alternate text for images and ensuring content is well-structured and navigable, also enhance SEO. Search engines favor websites that are accessible to all users, potentially increasing a site's visibility and traffic.

- **Reduced legal risk**: With the increasing global focus on digital accessibility, the risk of legal action against inaccessible websites is higher than ever. Investing in accessibility can protect businesses from costly litigation and the associated reputational damage.

- **Increased customer loyalty and brand perception**: Accessible websites demonstrate a company's commitment to inclusivity and social responsibility, improving brand perception and customer loyalty. Users who find a website accessible and easy to use are more likely to return and recommend the site to others.

- **Operational efficiencies**: Accessible design often leads to cleaner code and more streamlined website maintenance. By adhering to standard web practices required for accessibility, businesses can reduce the time and resources needed for website updates and troubleshooting.

Calculating the ROI of accessible web practices involves quantifying the direct and indirect benefits. Direct benefits can include increased sales from a broader customer base and reduced legal and compliance costs. Indirect benefits might encompass enhanced brand reputation, customer satisfaction, and operational efficiencies.

Organizations can assess ROI by tracking metrics such as increased traffic from users utilizing assistive technologies, higher conversion rates, and reduced maintenance costs over time. Furthermore, the avoidance of legal penalties and the value of positive brand perception, though harder to quantify, are significant factors that contribute to the overall ROI of investing in accessibility.

Demystifying the costs and benefits of accessible web practices reveals a compelling case for investment. The upfront costs associated with making websites accessible are outweighed by the extensive benefits, including expanded market reach, improved SEO, legal compliance, and enhanced brand loyalty. By viewing accessibility not as a cost but as an investment, businesses can realize significant returns that extend far beyond compliance and contribute to long-term success and sustainability.

Now that we understand accessibility to be an investment, let's explore the enhancements that it can make for users.

Enhancing UX and satisfaction

The enhancement of UX and satisfaction through accessible web practices stands as a pivotal argument for their adoption. This approach not only underscores the importance of inclusivity but also highlights how such measures significantly contribute to a positive ROI for businesses.

Accessible web design prioritizes usability, ensuring that websites are navigable, understandable, and operable for all users, including those with disabilities. This comprehensive approach to UX design considers various user needs, from those requiring screen readers to navigate content to individuals who rely on keyboard navigation. By implementing accessibility standards, businesses can enhance the overall UX, making it more seamless and enjoyable for a wider audience.

Websites that are accessible tend to have higher levels of user satisfaction. This satisfaction stems from ease of use, reduced barriers to access, and the positive perception of a brand that values all customers' experiences. Satisfied users are more likely to engage deeply with the website, spending more time exploring content, utilizing services, and completing transactions. This increased engagement is a direct driver of ROI as it often leads to higher conversion rates, repeat visits, and enhanced customer loyalty.

In today's competitive digital landscape, exceptional customer service is a key differentiator for brands. Accessible web practices extend the concept of customer service into the digital realm, showing users that a business is committed to providing an equitable experience for everyone. This commitment can significantly enhance customer satisfaction and perception of the brand, leading to strong customer relationships and advocacy.

Businesses can measure the impact of enhanced UX and satisfaction on their ROI through various metrics, including reduced bounce rates, increased time on site, higher conversion rates, and positive feedback through user surveys. Additionally, analytics can reveal how accessibility improvements lead to broader market reach and engagement, further contributing to a business's bottom line.

Enhancing UX and satisfaction through accessible web practices is not just a matter of compliance or ethical responsibility; it's a strategic business decision with a clear ROI. The improvements in usability and accessibility lead to higher user engagement, customer satisfaction, and brand loyalty, all of which contribute to the financial and reputational growth of a business. As such, investing in accessible web design is investing in the broader success and sustainability of a business in the digital age.

Let's break down some of the different ways you can broaden your market reach.

Broadening market reach

Accessibility is not just a conduit for inclusivity but also a powerful tool for expanding a company's audience. This section delves into how making web content accessible to people with disabilities can unlock untapped market potential, providing a significant ROI by reaching wider demographics.

Let's dive into specifics.

Expanding the audience base

Accessible web design inherently breaks down barriers to content consumption, making information, services, and products available to a broader audience, including the millions worldwide with disabilities. This inclusivity directly translates to an expanded customer base, encompassing users who might otherwise be excluded due to inaccessible web practices. By adhering to accessibility standards, businesses can cater to a diverse range of needs and preferences, thereby tapping into new segments of the market.

Enhancing global reach

Accessibility has global implications. As the internet becomes increasingly borderless, the potential to reach international audiences grows. However, this global reach necessitates a web presence that accommodates diverse abilities, including those affected by language barriers, varying degrees of internet literacy, and disabilities. Accessible websites, with features such as alternative text for images, subtitles for videos, and navigable structures for screen readers, ensure that businesses can engage with a global audience more effectively.

Increasing spending power

The disability market represents a significant and often overlooked spending power. In the US alone, the disposable income of people with disabilities and their families runs into billions of dollars. By making web content accessible, businesses can tap into this spending power, converting previously marginalized groups into loyal customers. The ROI in accessibility can be measured not just in terms of enhanced user engagement and satisfaction but also in the direct financial impact of accessing this sizable market.

Leveraging legal compliance for market advantage

In many regions, web accessibility is not merely a best practice but a legal requirement. While compliance avoids potential legal repercussions, it also presents an opportunity to outpace competitors who may be slower to adopt accessible web practices. Businesses that proactively embrace accessibility can market themselves as inclusive brands that appeal to a wider, socially conscious consumer base that values corporate responsibility.

Case studies highlighting market expansion

Numerous enterprises have documented significant growth following the adoption of accessible web practices. For instance, a leading e-commerce platform reported an increase in customer retention and a broader customer demographic after revamping its site to meet accessibility guidelines. Another case involved a multinational corporation that saw its market share increase in regions with high populations of users with disabilities, attributing this success to its accessible web design.

The broadening of market reach through accessible web practices underscores a vital business strategy in today's digital economy. Accessibility is not merely an ethical imperative but a business-savvy approach to inclusively and effectively engaging with a wider audience. By investing in accessible web design, businesses not only ensure compliance with legal standards but also open doors to untapped markets and demographics, securing a competitive edge and a robust ROI.

Legal compliance and risk mitigation

The imperative of legal compliance and risk mitigation emerges as a critical consideration for businesses operating in the digital realm. This section elucidates how adhering to web accessibility standards not only safeguards businesses from legal risks but also enhances their ROI by fostering a proactive culture of inclusivity and compliance.

Navigating the legal landscape

The legal landscape regarding web accessibility has become increasingly complex and enforceable across various jurisdictions worldwide. Laws and regulations such as the **Americans with Disabilities Act (ADA)** in the US, the **Accessibility for Ontarians with Disabilities Act (AODA)** in Canada, and the European Union's Web Accessibility Directive mandate that digital content must be accessible to users with disabilities. Non-compliance with these standards exposes businesses to legal challenges, including lawsuits, fines, and reputational damage.

The cost of non-compliance

The repercussions of failing to meet accessibility standards extend beyond the immediate financial penalties. Legal actions related to accessibility non-compliance can result in substantial costs, including legal fees, settlement amounts, and the expenses associated with retrofitting websites to meet accessibility guidelines. Moreover, the negative publicity associated with such legal challenges can erode customer trust and loyalty, impacting long-term revenue and growth.

Risk mitigation through accessibility

Investing in accessible web practices is a proactive measure that mitigates the risk of legal complications. By ensuring that websites and digital content are designed and developed to be accessible from the outset, businesses can avoid the costly process of addressing accessibility in response to legal pressures. This proactive approach not only reduces the risk of non-compliance but also positions businesses as leaders in CSR, enhancing brand reputation.

ROI of legal compliance

The ROI in achieving legal compliance extends beyond risk mitigation. Accessible web design improves the overall UX, potentially increasing site traffic, user engagement, and conversion rates. Additionally, the process of aligning with accessibility standards can uncover opportunities for improving website performance and efficiency, further contributing to the business's bottom line.

Building a culture of inclusivity

Beyond the technical and legal aspects of web accessibility, there's a significant opportunity for businesses to cultivate a culture of inclusivity. This culture values diversity and recognizes the importance of providing equitable access to digital resources for all users, including those with disabilities. Companies that champion accessibility can attract a diverse workforce and customer base, driving innovation and capturing broader market segments.

Legal compliance and risk mitigation in the context of web accessibility are not merely regulatory hurdles but strategic business imperatives. By embracing accessible web practices, businesses can protect themselves against legal risks, enhance their market competitiveness, and affirm their commitment to social responsibility. Therefore, investing in accessibility is not just a cost of doing business but a savvy strategy that yields significant returns through increased user engagement, brand strengthening, and compliance-driven innovation.

Summary

In summarizing the insights from this chapter, we have journeyed through the landscape of web accessibility, uncovering its profound impact beyond the realm of legal compliance. This chapter has equipped you with a comprehensive understanding of how embracing accessible web practices offers a multitude of benefits for businesses, including expanding market reach, enhancing user satisfaction, and fostering a synergy between RWD and accessibility.

We've learned that accessible design is not just about meeting the needs of users with disabilities; it's about creating a more inclusive digital space that benefits all users. By integrating accessibility principles into their digital strategy, businesses can tap into untapped market segments, improve their brand's reputation, and achieve a competitive edge in the digital marketplace. The case studies presented have illustrated the tangible returns businesses can achieve by prioritizing accessibility, from increased engagement and customer loyalty to significant ROI and legal risk mitigation.

These lessons are invaluable for any organization looking to thrive in today's digital landscape. Understanding the broader benefits of accessibility, the role of inclusive design in expanding market reach, and the interplay between RWD and accessibility equips businesses with the knowledge to make informed decisions that align with both their ethical standards and strategic goals.

As we move forward to the next chapter, we'll build on the foundation that's been laid by discussing web accessibility. This next chapter will delve into how accessible and responsive design practices not only benefit users but also play a crucial role in SEO. We'll explore how making your website more accessible can enhance its visibility on search engines, drawing a direct line between the technical aspects of accessibility and SEO performance. This natural progression underscores the interconnectedness of accessibility, UX, and SEO, highlighting how these elements collectively contribute to a successful digital presence.

9

SEO Considerations in Responsive Design

In the rapidly evolving digital landscape, the importance of aligning **RWD** with **search engine optimization** (**SEO**) strategies has never been more critical. This chapter delves into the intricate relationship between mobile search behavior, Google's mobile-first indexing approach, and the unique SEO challenges that come with RWD. This chapter aims to equip you with the necessary insights to not only navigate but also excel in optimizing your website for the best possible performance in search engine results.

As the internet becomes increasingly mobile-dominated, understanding the nuances of mobile search behavior is paramount. This chapter explores current trends, highlighting how users interact with search engines on mobile devices and the implications for businesses aiming to capture this audience. With Google's shift to mobile-first indexing, the stakes for having a mobile-optimized website are at an all-time high. This chapter breaks down what mobile-first indexing means for webmasters and SEO professionals, emphasizing the need for websites that perform excellently across all device types.

Moreover, this chapter addresses common SEO challenges inherent in RWD, from ensuring content parity between mobile and desktop versions to optimizing site speed and enhancing usability for mobile users. Through a comprehensive exploration of these topics, you will gain actionable insights and strategies to ensure your website is not only responsive but also fully optimized to meet the demands of modern search engines and users alike.

By the end of this chapter, you will have a well-rounded understanding of the critical role SEO plays in the success of responsive websites. You will be equipped with the knowledge to implement best practices in RWD, ensuring your site achieves and maintains high visibility in search results, provides an outstanding user experience, and ultimately drives engagement and conversions in a mobile-centric world.

The following topics will be covered in this chapter:

- The mobile search landscape
- Google's emphasis on mobile-first indexing
- RWD and its SEO advantages
- Avoiding common responsive SEO pitfalls

The mobile search landscape

The digital landscape has witnessed a monumental shift over the past two decades, largely due to the evolution of mobile usage in web searches. This transformation has redefined how users interact with the internet, prompting businesses and web developers to rethink their digital strategies to cater to the mobile-first world. Understanding this evolution is crucial for optimizing RWD in alignment with SEO considerations.

In the early days of the internet, desktop computers were the primary gateway to the web. Users would sit at their desks to browse websites, shop online, or access information. So, the design and optimization of websites were desktop-centric, focusing on larger screens and mouse-based navigation. However, the introduction and subsequent explosion of smartphones and tablets have dramatically altered this scenario.

The advent of the iPhone in 2007 marked a pivotal moment, setting the stage for the smartphone revolution. As mobile devices became more sophisticated and internet speeds increased, users gradually shifted from stationary desktop browsing to on-the-go web access. This shift was not merely a change in device preference but represented a fundamental change in user behavior, expectations, and interaction with digital content.

Mobile devices have introduced the convenience of accessing information anytime, anywhere. This has led to an increase in micro-moments, where users turn to their devices for quick answers to immediate questions. Consequently, search queries have become more focused, local, and intent-driven. Users expect fast, relevant answers to their location-based queries, such as finding nearby restaurants, services, or shops, which has significantly influenced how search engines such as Google rank and present search results.

In response to this shift, Google announced mobile-first indexing, a strategy that prioritizes the mobile version of a website's content for indexing and ranking. This move underscores the importance of mobile-friendly web design and content optimization as Google aims to serve its predominantly mobile user base more effectively. Websites that offer a superior mobile experience are now favored in search results, highlighting the necessity for RWD, which ensures content is accessible and engaging across all device types.

The evolution of mobile usage has direct implications for SEO. It necessitates an approach that goes beyond mere mobile compatibility, demanding that websites be designed with the mobile user's needs and behaviors in mind. This includes considerations for page speed, site design, and content layout that align with the way users search and consume content on mobile devices. As the mobile web continues to grow, staying abreast of these changes and optimizing for mobile search behavior is paramount for staying in the lead.

Let's explore the impact that mobile searches have had on users.

The impact of mobile searches on user behavior

The profound impact of mobile searches on user behavior underscores a significant shift in how information is consumed and acted upon in the digital age. As smartphones have become ubiquitous, the immediacy and convenience they offer have reshaped user expectations and interaction patterns with the web. This transformation is pivotal for understanding the nuances of SEO in a mobile-dominated landscape and optimizing RWD accordingly.

Mobile devices facilitate access to information at an unprecedented speed, fostering an environment where users expect immediate results. This expectation of instantaneity influences not only the types of searches conducted but also how users engage with search results. Users are more likely to favor websites that load quickly and present the sought-after information in an easily digestible format, preferably at the top of the page or within the first few scrolls.

The portability of mobile devices has led to a surge in localized searches, with users often seeking information relevant to their immediate geographical location. Queries such as *near me* or *open now* have become commonplace, prompting search engines to prioritize local business listings and geographically relevant content in their results. This trend necessitates that businesses optimize their online presence for local SEO, ensuring their sites are visible and appealing to potential customers in their vicinity.

Voice search technology, powered by AI assistants such as Siri, Google Assistant, and Alexa, has introduced a new dimension to mobile searches. Users are increasingly relying on voice commands to conduct searches, leading to more conversational, long-tail queries. This shift toward natural language searches requires a nuanced approach to keyword optimization, emphasizing semantic search and content that answers direct questions or solves specific problems.

The capabilities of modern smartphones, combined with the visual nature of mobile web browsing, have propelled the popularity of visual and video content in search queries. Platforms such as Instagram and YouTube, along with features such as Google Lens, cater to users' preferences for visual information, influencing how businesses should present their content. Optimizing visual and video content for searchability — through descriptive titles, tags, and alternative text — becomes crucial in attracting and retaining mobile users.

Understanding the impact of mobile searches on user behavior is essential for tailoring SEO strategies to meet these evolving needs. Websites optimized for mobile must not only be technically sound, loading quickly and functioning properly across devices but also content-rich, offering valuable information that aligns with users' intent and preferences. The integration of local SEO, voice search optimization, and the strategic use of visual content are key components in enhancing a website's visibility and engagement in a mobile-first world.

To summarize, the impact of mobile searches on user behavior dictates a comprehensive approach to SEO that transcends traditional tactics. Embracing the nuances of how users search on mobile and optimizing the web experience to cater to these behaviors is paramount for businesses aiming to thrive in the competitive digital ecosystem.

Now, let's break down how the largest company in the internet space approaches web indexing.

Google's mobile-first approach to web indexing

Google's mobile-first approach to web indexing marks a pivotal shift in the search engine's strategy, fundamentally changing how websites are crawled, indexed, and ranked. This evolution is a direct response to the increasing dominance of mobile web usage, reflecting Google's commitment to providing the most relevant and user-friendly search results. Understanding this approach is crucial for SEO professionals and web developers aiming to optimize their sites for visibility and performance.

Historically, Google's indexing system primarily used the desktop version of a website's content to evaluate its relevance to a user's query. However, with the majority of users now accessing Google via mobile devices, this method has become outdated. Mobile-first indexing means Google predominantly uses the mobile version of the content for indexing and ranking. This transition underscores the necessity of having a mobile-responsive site not just for user experience but also as a fundamental component of the SEO strategy.

The shift to mobile-first indexing requires developers to ensure their mobile sites are as comprehensive and optimized as their desktop versions. Sites that employ RWD are in the best position as their content is identical across devices. However, for sites with separate mobile versions, it's vital to ensure content parity — meaning the mobile site should offer the same quality and quantity of content as its desktop counterpart. This includes text, images, videos, and links. Mobile sites should also be structured so that Googlebot can crawl them efficiently, with metadata and structured data being consistent across both versions.

To align with Google's mobile-first approach, several key areas require attention:

- **Responsive design**: Ensuring your site is responsive is more critical than ever. A responsive site automatically adjusts its layout and content to fit the screen it's being viewed on, providing an optimal user experience.
- **Site speed**: Mobile users expect quick loading times. Optimizing images, leveraging browser caching, and minimizing code are effective strategies to improve site speed.

- **User experience**: Beyond aesthetics, the mobile site's u should facilitate easy navigation and interaction, with touch-friendly menus and buttons, readable fonts, and accessible design elements.

- **Content optimization**: Content should be optimized for mobile consumption while considering shorter attention spans and the propensity for scrolling. Clear headings, bullet points, and concise language can enhance readability and engagement.

Embracing Google's mobile-first indexing is not a one-time task but an ongoing process of monitoring and adaptation. Site owners should regularly use Google's Search Console to monitor mobile usability issues and address them promptly. Staying informed about updates to Google's algorithms and mobile-first indexing practices will enable SEO professionals to adjust their strategies, ensuring their sites remain competitive and visible in mobile search results.

The adoption of Google's mobile-first approach to web indexing signifies a broader recognition of the evolving ways users interact with the web. For businesses and developers, this shift reinforces the importance of adopting a mobile-centric approach to web design and SEO, ensuring that sites are not only discoverable but also deliver a satisfying user experience on the smallest screens.

Let's explore why Google emphasizes mobile-first indexing.

Google's emphasis on mobile-first indexing

Google's mobile-first indexing represents a fundamental shift in how the world's largest search engine processes and ranks websites. This change underscores the increasing prevalence of mobile browsing, reflecting a broader trend in user behavior. As the majority of internet users now access the web via mobile devices, Google has adjusted its indexing strategy to prioritize mobile versions of content. Understanding this shift is crucial for anyone involved in SEO or web design as it directly impacts how sites are perceived and ranked by Google.

Mobile-first indexing means that Google predominantly uses the mobile version of a website's content for indexing and ranking. Historically, Google indexed the desktop version of a website first, but with mobile-first indexing, the mobile version becomes the starting point for what Google includes in its index and the baseline for how it determines rankings. This shift doesn't mean Google has a separate index for mobile content; it simply prioritizes mobile content over desktop content in its existing index.

The shift to mobile-first indexing is Google's response to the evolution of user access patterns. With more users now accessing the web on mobile devices than on desktop computers, the old model of prioritizing desktop content no longer serves the majority of users' needs. Mobile-first indexing seeks to improve the user experience by ensuring that users find content that is optimized for the small screen and touch-based navigation of their primary browsing devices.

Let's explore some of the implications.

Implications for site owners and SEO professionals

For webmasters and SEO professionals, mobile-first indexing means that having a mobile-optimized or responsive site is more critical than ever. Here are some of the key implications:

- **Content parity**: Ensuring that mobile versions of sites have the same quality and quantity of content as their desktop counterparts is vital. This includes text, images, videos, and links. Anything less can negatively impact a site's ranking in search results.

- **Structured data**: Structured data should be present on both versions of the site.

- **Search console verification**: If a site has separate mobile and desktop versions, Google recommends verifying both in Search Console to monitor crawl rates and indexing.

The next question to answer would be, *What are the best practices for mobile-first indexing?*

Best practices for mobile-first indexing

Adapting mobile-first indexing involves several best practices:

- **Responsive design**: Employing RWD ensures that the same content and structure are presented across devices, adapting seamlessly to any screen size

- **Speed optimization**: Mobile pages should load quickly as site speed is a significant factor in mobile browsing satisfaction and, consequently, SEO rankings

- **User experience**: The mobile user experience should be a primary focus, with easily navigable menus, accessible content, and touch-friendly elements

Understanding and adapting to mobile-first indexing is not just about appeasing Google's algorithms — it's about aligning with the natural progression of the web toward mobile. By prioritizing the mobile experience in SEO strategies and web design, businesses and webmasters can ensure their sites meet the needs of today's users, thereby securing their visibility and competitiveness in the digital landscape.

Now, let's explore how we can prepare for mobile-first indexing.

Preparing for mobile-first indexing

Regularly testing a website on various mobile devices and browsers is essential to identify and resolve usability issues that could affect mobile-first indexing. Tools such as Google's Mobile-Friendly Test can assess how easily a user can access a page on a mobile device. Additionally, gathering user feedback on the mobile experience can offer actionable insights for further optimization.

A mobile-optimized site should prioritize user experience by ensuring that text is readable without zooming, site elements are easily clickable, and the overall layout is navigable on a small screen. Accessibility should also be considered, making sure that content is accessible to all users, including those with disabilities.

Preparing for mobile-first indexing is not just about making a site mobile-friendly but about embracing a mobile-first philosophy in all aspects of website design and content strategy. By prioritizing the mobile experience, websites can not only align with Google's indexing practices but also meet the evolving expectations of modern web users.

Let's look at how SEO has been impacted by mobile-first indexing.

The impact of mobile-first indexing on SEO strategies

Google's shift toward mobile-first indexing has profound implications for SEO strategies, compelling digital marketers and webmasters to reassess and realign their approaches to cater to a predominantly mobile audience. This paradigm shift underscores the importance of prioritizing the mobile user experience as the mobile version of a website now serves as the primary source for Google's index and ranking algorithms. Understanding the impact of this change is crucial for maintaining and enhancing a website's visibility in search results.

With mobile-first indexing, the usability of the mobile version of a website becomes a critical factor in its search performance. SEO strategies must now place a greater emphasis on mobile design principles, ensuring that sites are intuitive, navigable, and accessible on smaller screens. This includes optimizing button sizes, ensuring text readability without zooming, and streamlining menus and navigation to improve the ease of use on mobile devices.

Page loading speed has always been a factor in SEO, but with mobile-first indexing, its importance is magnified due to the variability of mobile internet speeds and the typically lower processing power of mobile devices compared to desktops. SEO strategies must incorporate techniques for speed optimization, such as image compression, minimizing code, leveraging browser caching, and using **accelerated mobile pages** (**AMPs**) to ensure swift loading times on mobile devices.

SEO content strategies need to adapt to the consumption habits of mobile users, who typically seek quick, easily digestible information. This necessitates a shift toward more concise, impactful content, optimized for mobile consumption. The structure of content may also need adjustment, with more frequent use of bullet points, shorter paragraphs, and engaging headers to facilitate skimming and scanning.

The prevalence of mobile searches has led to a surge in local search queries as users look for businesses, services, and products in their immediate vicinity. This trend makes local SEO more crucial than ever, requiring businesses to optimize their online presence for local search by ensuring their Google My Business listings are up to date, incorporating local keywords into their SEO strategy, and generating local content.

The rise of voice search and natural language queries, driven by mobile device usage, requires reevaluating keyword strategies. Long-tail keywords and conversational phrases that mirror spoken language are increasingly important for SEO. Optimizing for these types of queries means adopting a more natural, user-focused approach to content creation.

Structured data markup helps search engines understand the content and context of a website, facilitating the presentation of more informative and visually appealing results in **search engine results pages** (**SERPs**). With mobile-first indexing, the implementation of structured data becomes even more critical as it can enhance visibility and click-through rates from mobile search results.

While mobile-first indexing prioritizes the mobile version of a website, it's essential to remember that this does not mean mobile-only. A holistic approach to SEO that ensures an outstanding user experience across all devices is still necessary. A responsive design that adjusts seamlessly between desktop and mobile, without sacrificing content quality or accessibility, remains the gold standard.

The impact of mobile-first indexing on SEO strategies is significant, requiring a shift in focus toward optimizing all aspects of a website's mobile presence. By embracing these changes and prioritizing the mobile user experience, businesses can align themselves with Google's indexing practices and the preferences of the modern web user, securing their place in the competitive digital landscape.

Now, let's explore the impact of SEO on RWD.

RWD and its SEO advantages

RWD is not only a cornerstone of modern web development but also Google's recommended approach to creating mobile-friendly websites. This endorsement by the world's leading search engine underscores the significant SEO advantages that RWD offers, aligning with the shift toward mobile-first indexing and the ever-increasing emphasis on user experience. Understanding why Google advocates for RWD can provide valuable insights into developing effective SEO strategies that cater to the demands of today's digital landscape.

Google's preference for RWD is closely tied to its mobile-first indexing initiative. As the indexing process primarily uses the mobile version of a website's content to determine its search ranking, RWD ensures that the same content is accessible across all devices without the need for separate URLs or differing HTML for mobile and desktop versions. This uniformity simplifies Google's crawling and indexing process, making it easier for the search engine to understand and accurately rank content.

At the heart of RWD's SEO advantages is its ability to provide an optimal viewing experience — easy reading and navigation with minimal resizing, panning, and scrolling — across a wide range of devices. Google has long factored user experience into its ranking algorithms, and with the increasing use of mobile devices to access the internet, a site's mobile user experience is paramount. RWD directly improves user satisfaction, which can lead to lower bounce rates, higher engagement rates, and, ultimately, better search rankings.

Google's recommendation of RWD also stems from the operational efficiencies it offers to website owners. Managing one responsive website rather than separate sites for mobile and desktop users reduces the time and resources required for site maintenance. This singular approach eliminates common discrepancies between mobile and desktop content, ensuring consistency in SEO efforts and messaging across all platforms. Furthermore, it streamlines the process of implementing and updating SEO strategies as changes only need to be made once.

Speed is a critical factor in both user experience and SEO, particularly on mobile devices where users expect quick loading times. Responsive designs often incorporate best practices for optimizing speed, such as image compression, efficient code, and responsive image solutions, which can significantly reduce page load times. Faster sites not only rank better on Google but also retain users more effectively, contributing to higher conversion rates.

RWD facilitates easier social sharing by providing a consistent URL structure across devices. When users share content from a responsive site, the link opens seamlessly on any device, enhancing the content's reach and engagement — factors that indirectly influence search rankings through increased traffic and user interaction signals.

Google's recommended approach being RWD is a clear indicator of the direction in which web development and SEO are heading. By prioritizing RWD in their digital strategies, businesses can align with Google's best practices for search optimization, ensuring their sites are not only more discoverable but also more engaging for users regardless of how they access the web. This alignment between user experience and SEO best practices underscores the integral role RWD plays in the success of modern websites in achieving both their user engagement and search visibility goals.

Now, it's time to break down how to boost our site.

Boosting site usability and user engagement

RWD serves as a critical tool in enhancing site usability and user engagement, two factors that are increasingly vital in the realm of SEO. As search engines evolve to prioritize the user experience, the role of RWD in creating websites that are both accessible and engaging across all devices has never been more essential. This approach not only meets users' expectations for seamless browsing but also aligns with search engines' criteria for determining site quality and relevance.

RWD eliminates the need for separate site designs for desktops, tablets, and smartphones by using fluid grids, flexible images, and CSS media queries to adapt the layout of a website to the viewing environment. This adaptability improves site usability by ensuring that users have a consistent experience, regardless of the device they use. Navigation menus, content, and interactive elements automatically adjust to the screen size, minimizing zooming, panning, and scrolling, which can detract from the user experience and lead to higher bounce rates.

Websites that are easy to use and navigate are more likely to keep visitors engaged. RWD contributes to a positive user experience by providing quick access to information, simplifying the path to conversion, and ensuring that interactive elements work correctly on touchscreens. By improving the ease of interaction, responsive websites can significantly increase time on site and page views, which are critical metrics for SEO success. Engaged users are more likely to consume more content, interact with calls to action, and become loyal visitors or customers.

A key benefit of RWD is its potential to reduce bounce rates. When users encounter a site that loads quickly and displays content appropriately for their device, they are less likely to leave the site without engaging further. Lower bounce rates signal to search engines that a website is providing valuable content that meets users' needs, which can positively impact search rankings.

The ultimate goal of many websites is to convert visitors into leads or customers. RWD plays a pivotal role in this process by ensuring that forms, buttons, and other conversion elements are easily accessible and usable on mobile devices. Optimizing these elements for touch interaction and making sure they are prominently displayed can lead to higher conversion rates, directly affecting a site's ROI and indirectly benefiting its SEO by demonstrating value to users.

With mobile internet usage surpassing desktop, the ability of RWD to cater effectively to mobile users is a significant SEO advantage. Google and other search engines recognize mobile-friendly sites in their rankings, often giving them precedence over non-responsive sites for searches conducted on mobile devices. Furthermore, as voice search and local searches become increasingly prevalent on mobile, responsive sites that offer an optimized experience are better positioned to capture this traffic.

The direct correlation between site usability, user engagement, and SEO rankings underscores the importance of implementing RWD. By enhancing the user experience across all devices, RWD not only meets the technical requirements of mobile-first indexing but also addresses the broader goal of satisfying user intent. In the competitive landscape of search engine rankings, the ability of RWD to boost site usability and user engagement represents a powerful strategy for achieving and maintaining high visibility in search results.

Next, we'll explore how RWD helps with sharing and distribution.

Enhancing social sharing and content distribution

RWD significantly enhances social sharing and content distribution, making it easier for users to share content across different platforms and devices. This seamless sharing capability is crucial for broadening a website's reach and can indirectly influence its SEO performance by driving more traffic back to the site. As social signals continue to play a role in how search engines understand and rank web content, the importance of optimizing for social sharing within RWD cannot be overstated.

RWD eliminates the barriers that users often face when sharing content between different types of devices. Whether users are engaging with content on a smartphone, tablet, or desktop, RWD ensures that shared links open correctly and display content in an optimized format across all devices. This uniformity improves the user experience for content recipients, increasing the likelihood of further engagement and sharing.

One of the key benefits of RWD in the context of social sharing is the use of a single URL for a piece of content, regardless of the device it's viewed on. This consistency simplifies the sharing process as there's no need to manage separate URLs for mobile and desktop versions of a site. Unified URLs help maintain link equity — a factor in SEO rankings — and ensure that all social shares contribute to the overall popularity and visibility of the content.

Content that looks good and is easy to interact with is more likely to be shared. RWD facilitates the creation of visually appealing and accessible content across all screen sizes, enhancing engagement. Well-designed responsive sites can adjust images, fonts, and layout elements dynamically, ensuring that content is not only readable but also engaging on any device. This attention to detail can significantly increase the content's shareability, attracting more views and interactions.

The flexibility of RWD allows for the effective presentation of various content formats, including text, images, videos, and infographics. By accommodating these different types of media, responsive sites can cater to diverse user preferences, encouraging the distribution of a wider range of content. Optimizing videos and images for fast loading times and high-quality display on mobile devices further enhances the sharing experience, contributing to increased social traffic.

Social sharing acts as a powerful amplifier for content, extending its reach beyond a website's immediate audience. Each share introduces content to a new network of potential viewers, increasing its visibility and driving additional traffic. This social validation not only contributes to brand awareness but can also positively impact SEO by generating more inbound links and mentions, which are important signals to search engines.

In the digital marketing ecosystem, the ability to share content easily and effectively is paramount. RWD enhances social sharing and content distribution by ensuring that content looks and performs optimally across all devices. This optimization not only improves the user experience but also amplifies the reach of content through social networks, indirectly supporting SEO efforts. As social interactions continue to influence online visibility, the role of RWD in facilitating these interactions becomes increasingly critical, making it an essential component of any effective SEO strategy.

Now, let's look at some of the common SEO pitfalls that we should avoid.

Avoiding common responsive SEO pitfalls

Navigating the landscape of RWD requires a keen understanding of common pitfalls that can hinder a site's SEO performance. Among these, misconfigured viewports and designs that are not mobile-friendly stand out as critical issues that can significantly impact a site's usability and, by extension, its ranking in search results. Addressing these pitfalls is essential for ensuring that a site is fully optimized for both users and search engines.

Let's take a closer look at these pitfalls, starting with the next topic: *Misconfigured viewport*.

Misconfigured viewport

The viewport is the user's visible area of a web page, and it varies with the device. Configuring the viewport properly is crucial for RWD because it instructs the browser on how to adjust the page's dimensions and scaling to suit the viewing device. A misconfigured viewport can result in content that is too small to read, requires horizontal scrolling, or is otherwise difficult to navigate on a mobile device. Such issues not only frustrate users but can also negatively affect a site's mobile usability score – a factor that Google considers in its ranking algorithm.

To avoid this pitfall, it's important to use the viewport meta tag in the head of the HTML document to control the layout on mobile browsers effectively. This tag should specify the width of the viewport so that it matches the device's width, and the initial scale should be set to 1.0 to ensure that the page is displayed at the appropriate size.

Unfriendly mobile design

Even with a configured viewport, a design can still be unfriendly to mobile users if it fails to consider the nuances of mobile navigation and interaction. The following are some common indicators of unfriendly mobile design:

- **Tiny fonts that are hard to read on smaller screens**: Ensuring that text sizes are responsive and easily legible without zooming is fundamental for mobile usability

- **Inaccessible menus and navigation links**: Menus should be easily expandable and clickable, with adequate spacing around links to prevent mis-taps

- **Slow-to-load images and media**: Optimizing images and media files for fast loading on mobile networks improves the user experience and contributes to higher engagement rates

- **Lack of touch-friendly elements**: Design elements should be sized and spaced so that they can be easily tapped with a finger without the risk of the user selecting the wrong link or button

Addressing these design issues requires a thoughtful approach that prioritizes the mobile user's experience. Techniques such as using flexible grids and images, prioritizing content hierarchy, and implementing touch-friendly navigation menus can transform a site from being merely mobile-compatible to genuinely mobile-optimized.

SEO implications

Google's mobile-first indexing and the increasing importance of mobile usability metrics mean that misconfigured viewports and unfriendly mobile designs can have a direct impact on SEO. Sites that provide a suboptimal mobile experience may see a decrease in their search rankings, especially for searches conducted on mobile devices. Conversely, sites that embrace best practices for mobile design are more likely to rank well and attract more traffic.

In summary, avoiding the pitfalls of misconfigured viewports and unfriendly mobile design is not just about adhering to technical best practices; it's about embracing the mobile-first philosophy that underpins modern web usage. By ensuring that responsive designs meet the needs of mobile users, site owners, and SEO professionals can enhance both the user experience and the site's visibility in search engine results.

Slow mobile page speed

Slow mobile page speed is a critical issue that can severely impact a site's user experience and search engine rankings. In an era where speed is equated with efficiency, mobile users have little patience for slow-loading pages, and search engines such as Google have incorporated page speed into their ranking algorithms. Understanding the implications of slow mobile page speed and adopting strategies to mitigate this issue is essential for any SEO-friendly RWD.

The impact of slow mobile page speed

Mobile users often browse the web in scenarios where they expect quick information access, such as comparing prices while shopping or looking up directions. A delay of just a few seconds can lead to frustration, increased bounce rates, and lost opportunities for engagement and conversion. Moreover, Google's mobile-first indexing makes mobile page speed a critical factor in SEO. Sites that load quickly on mobile devices are likely to be favored in search results, making speed optimization a priority for webmasters and SEO professionals.

Identifying the causes of slowness

Several factors can contribute to slow mobile page speeds, including the following:

- **Unoptimized images**: Large images that are not compressed for the web can significantly slow down page loading times.

- **Excessive JavaScript and CSS**: Bloated code or excessive use of JavaScript and CSS can lead to longer processing times, particularly on mobile devices with less computing power than desktops.

- **Server response time**: The amount of time it takes for a server to respond to a request can add to overall page load times. This can be influenced by the server's location, configuration, and the quality of the hosting service.

- **Render-blocking resources**: Resources that must be loaded before the page can be displayed, such as certain JavaScript or CSS files, can delay the rendering process.

Now, let's consider some strategies for speed optimization.

Strategies for speed optimization

Improving mobile page speed requires a multi-faceted approach that focuses on both the size and complexity of the resources that have been loaded on a page:

- **Optimize images**: Use compression tools to reduce image file sizes without sacrificing quality and consider implementing responsive images that adjust to screen sizes.

- **Minimize code**: Minify CSS, JavaScript, and HTML by removing unnecessary characters, spaces, and comments. Use tools that automatically streamline code to enhance performance.

- **Leverage browser caching**: Configure your server to enable caching for repeat visitors, reducing the amount of data that needs to be transferred over the network.

- **Eliminate render-blocking resources**: Move scripts to the bottom of the HTML or use asynchronous loading to allow the rest of the page to load without having to wait for these resources.

- **Use a content delivery network (CDN)**: CDNs distribute your content across multiple servers around the world, reducing the distance between users and server resources, which can speed up loading times.

- **Optimize server response time**: Choose a reliable hosting solution and consider using technologies such as HTTP/2 to improve the efficiency of content delivery.

Next, we'll look at monitoring and continuous improvement.

Monitoring and continuous improvement

Regularly monitoring mobile page speed using tools such as Google's PageSpeed Insights can provide actionable recommendations for improvement. It's important to treat speed optimization as an ongoing process, continually seeking ways to enhance site performance as user expectations evolve and new optimization techniques emerge.

In conclusion, addressing slow mobile page speed is not only crucial for providing a positive user experience but also for improving a site's visibility and rankings in search engine results. By implementing targeted optimization strategies, websites can improve their responsiveness, engage users more effectively, and achieve better SEO outcomes in the mobile-first digital landscape.

Content hidden in tabs or accordions

The use of tabs, accordions, and other collapsible content formats in RWD is a common practice that aims to improve the user experience on mobile devices, as well as some desktops. These elements help manage and organize content efficiently, making it more accessible and navigable for users on smaller screens. However, from an SEO perspective, historically, there's a nuanced consideration to be mindful of: content hidden in tabs or accordions was weighted differently by search engines, potentially impacting a site's visibility and rankings.

The evolution of the search engine approach

In the past, search engines such as Google often gave less weight to content hidden in collapsible tabs or accordions. The rationale was that if the content is not immediately visible to the user, it might not be as important or relevant. However, as mobile usage soared and RWD became more prevalent, Google adapted its approach. Recognizing that these design elements are used to enhance usability on mobile devices, Google announced that content in tabs, accordions, or hidden behind **read more** buttons would be treated equally to visible content when indexing mobile sites. This shift acknowledged the balance between user experience and the need to present content efficiently on mobile devices.

Best practices for using collapsible content

Despite the evolution of search engines' treatment of hidden content, it's crucial to use collapsible elements wisely to ensure they don't inadvertently harm your SEO efforts. Here are some best practices:

- **Prioritize the user experience**: Use tabs and accordions to improve content organization and readability, particularly on mobile devices, without sacrificing accessibility or usability. Ensure that these elements function intuitively and enhance the overall user experience.

- **Maintain content quality and relevance**: The content within collapsible sections should be as high-quality and relevant as the rest of the page. It should provide value to the user and be directly related to the page's topic, utilizing appropriate keywords and answering user queries effectively.

- **Ensure accessibility**: Content hidden in collapsible elements should be accessible to all users, including those using screen readers or other assistive technologies. Proper use of HTML5 semantic elements, such as `<details>` and `<summary>`, can improve accessibility.

- **Avoid overuse**: While tabs and accordions can be useful, overusing them may lead to a cluttered and confusing user interface. Use these elements judiciously, keeping the most important information visible and using collapsible sections for supplementary content.

Finally, we'll cover SEO implications.

SEO implications

For SEO, the key is transparency and relevance. Search engines aim to serve users the most relevant content based on their queries. By ensuring that the content within tabs or accordions is directly relevant to the page's topic and user queries, webmasters can align with search engines' goal of providing value to users. Additionally, as mobile-first indexing becomes the standard, the way content is presented and organized for mobile users is increasingly significant for SEO.

The inclusion of content in tabs or accordions should be approached with a strategic mindset, balancing the need for a clean, navigable mobile experience with the imperative to make content accessible and valuable. By adhering to best practices for collapsible content and staying informed on search engines' evolving guidelines, webmasters and SEO professionals can leverage these design elements effectively, enhancing both user experience and search engine visibility.

Summary

As we conclude this chapter, we've traversed the complex terrain of optimizing websites in a mobile-first world. This journey has equipped us with vital insights into mobile search behaviors, underscored the significance of Google's mobile-first indexing, and navigated through the common yet intricate SEO challenges associated with RWD. Each section has been carefully crafted to not only impart knowledge but also to translate this knowledge into actionable strategies that can be applied to real-world scenarios.

The exploration of current mobile search trends has highlighted the pivotal role mobile devices play in today's digital ecosystem. Understanding these trends is crucial for tailoring SEO strategies that resonate with the mobile user's needs and search habits. Our deep dive into Google's mobile-first indexing has clarified its implications, emphasizing the necessity for websites to prioritize mobile optimization to maintain and enhance their search engine rankings.

Addressing typical SEO challenges in RWD, from content visibility issues in tabs or accordions to optimizing for speed and ensuring user-friendly navigation, has shed light on the multifaceted nature of SEO in the context of responsive design. These lessons underscore the importance of a holistic approach to RWD, one that seamlessly integrates SEO best practices to foster sites that are not only aesthetically pleasing and functional across devices but also optimized for peak performance in search engine results.

The skills and knowledge you've acquired in this chapter are invaluable for anyone involved in web design and development, offering the tools needed to create websites that excel in both form and function. In an increasingly competitive digital space, these insights can be the difference between a website that merely exists and one that thrives, attracting and engaging users while achieving superior search engine visibility.

Transitioning from the technical intricacies of SEO and RWD, the next chapter presents the next natural step in our exploration. Building on the foundations laid in understanding the technical and strategic aspects of responsive design, we'll shift our focus toward the leadership and management skills necessary to guide design teams toward success.

10
Leading and Guiding Design Teams

In this chapter, we delve into the nuanced and critical skills required for leadership in the dynamic field of **responsive web design (RWD)**. As digital landscapes evolve, so too must the leadership strategies that guide design teams toward success. This chapter is crafted to equip current and aspiring leaders with the knowledge and tools necessary to navigate the complexities of leading design teams in an era marked by rapid technological advancements and shifting user expectations.

Leadership in the digital age demands a unique blend of vision, adaptability, and a deep understanding of the technologies and methodologies driving change. This chapter explores how to develop these leadership qualities, emphasizing the importance of a mindset that embraces continuous learning and growth. We examine strategies for effectively communicating RWD principles, ensuring that teams not only understand the technical aspects of their work but also the broader business context and user-centered focus that make RWD so critical.

Fostering an environment of innovation within design teams is another cornerstone of effective leadership explored in this chapter. We delve into practical approaches for nurturing creativity, encouraging collaborative problem-solving, and implementing design thinking and Agile methodologies to adapt to challenges. By cultivating a culture that values experimentation and learning from failures, leaders can unlock the full creative potential of their teams.

Understanding metrics of success and the importance of continual refinement is also crucial. This chapter provides insights into setting benchmarks for RWD projects, employing tools and techniques for monitoring performance, and leveraging user feedback and behavior analytics to inform iterative improvements. Leaders will learn the value of A/B testing and creating a culture of continuous improvement, ensuring that design efforts are always aligned with user needs and business goals.

Through this comprehensive exploration, this chapter aims to empower leaders with the skills to not only navigate but thrive in the digital age. By mastering the art of conveying RWD concepts, fostering innovation, and measuring success, leaders can guide their teams to achieve exceptional results in creating responsive, user-centric digital experiences.

As we prepare to turn the page on this chapter, readers will be poised to apply these lessons in real-world contexts, leading design teams with confidence and insight. The journey ahead promises to be both challenging and rewarding as we continue to explore the ever-evolving landscape of RWD and digital leadership.

The topics covered in this chapter are the following:

- Leadership in the age of digital transformation
- Effective communication of RWD principles
- Measuring success and iterative improvements
- Nurturing creativity and problem-solving

Leadership in the age of digital transformation

In the realm of modern business, digital transformation has become a pivotal force driving innovation, efficiency, and growth. It's crucial to begin with a foundational understanding of what digital transformation entails and its profound impact on organizations and industries worldwide.

Digital transformation encompasses the integration of digital technology into all areas of a business, fundamentally changing how operations are conducted and how value is delivered to customers. It's not merely about adopting new technologies but involves a holistic rethinking of business models, processes, and strategies to leverage digital advancements for competitive advantage. This transformation extends beyond the company itself, influencing customer expectations, market dynamics, and the broader industry landscape.

Several key drivers propel organizations toward digital transformation. The exponential growth of data, advancements in technology such as cloud computing and **artificial intelligence** (**AI**), and the increasing digital savviness of consumers all play pivotal roles. Moreover, the rapid pace of digital innovation necessitates that businesses adapt not just to survive but to thrive and lead in their respective domains.

Leadership in the age of digital transformation is about more than guiding technological adoption; it's about cultivating a vision for the future and inspiring the organization to embrace change. Leaders must champion a culture of innovation, where experimentation is encouraged, and failures are viewed as stepping stones to success. They need to ensure that their teams are equipped with the necessary skills and mindset to navigate the digital landscape effectively.

For design teams, digital transformation presents both challenges and opportunities. The shift toward RWD and user-centered design principles is indicative of broader trends in digital transformation — prioritizing agility, **user experience** (**UX**), and data-driven decision-making. Leaders must guide their teams to understand and adopt these principles, ensuring that design practices align with the strategic goals of digital transformation.

Understanding digital transformation is the first step in preparing design teams for the continuous evolution of the digital ecosystem. Leaders must foster an environment where learning is continuous and adaptability is ingrained in the team's culture. By doing so, they can ensure that their teams are not just participants in the digital age but are at the forefront, driving innovation and shaping the future of their industries.

In conclusion, understanding digital transformation is foundational for leaders aiming to navigate their teams through the complexities of the modern digital landscape. By embracing the principles of digital transformation, leaders can empower their design teams to create responsive, user-centric solutions that not only meet but exceed the evolving expectations of users and stakeholders alike.

As leaders, it's our responsibility to help navigate change.

The role of leaders in navigating change

In the dynamic landscape of digital transformation, the role of leaders in navigating change is paramount. As organizations strive to adapt to the rapid advancements in technology and shifts in consumer behavior, effective leadership becomes the linchpin in steering teams toward success and innovation. This exploration focuses on the multifaceted responsibilities of leaders in guiding their design teams through the tumultuous waters of change inherent in the digital age.

At the heart of successful digital transformation is visionary leadership. Leaders must articulate a clear and compelling vision of how digital technologies can revolutionize the organization's operations, products, and services. This vision provides direction and purpose, aligning the team's efforts with the broader goals of digital transformation. It's not enough to respond to change reactively; leaders must anticipate future trends and prepare their teams to be at the forefront of innovation.

The fast-paced nature of digital transformation demands agility and resilience, both from leaders and their teams. Leaders must foster a culture where adaptability is valued, encouraging teams to pivot quickly in response to new information or market demands. This agility is complemented by resilience — the capacity to face setbacks and challenges without losing momentum. By modeling resilience and supporting their teams through failures and learning experiences, leaders can build a strong foundation for sustained innovation.

Digital transformation blurs the lines between traditional roles and departments, necessitating a higher degree of collaboration and cross-functional teamwork. Leaders play a critical role in breaking down silos and fostering an environment where designers, developers, marketers, and other stakeholders work seamlessly together. This synergy enhances the team's ability to deliver holistic, user-centered solutions that are responsive to the needs of a digitally connected audience.

Navigating change requires not just top-down directives but the empowerment of team members to make decisions and take ownership of projects. Leaders should delegate authority and provide teams with the autonomy to explore creative solutions. This empowerment boosts morale, encourages innovation, and accelerates the pace of digital transformation efforts. It also allows leaders to focus on strategic objectives, knowing that their teams are capable of handling operational challenges.

Perhaps the most potent tool in a leader's arsenal is leading by example. When leaders actively embrace change, show enthusiasm for new technologies, and demonstrate a willingness to learn, it sets a powerful precedent for the entire team. This behavior reinforces the importance of staying curious, being open to new ideas, and continuously evolving one's skills to remain relevant in a digital-first world.

The role of leaders in navigating change during digital transformation is multifaceted and impactful. By providing a clear vision, fostering a culture of agility and resilience, championing collaboration, empowering teams, and leading by example, leaders can guide their design teams through the challenges and opportunities of the digital age. This leadership approach ensures that teams are not only prepared to adapt to change but are also poised to drive innovation and lead the organization toward a successful digital future.

Embracing new technologies and methodologies

In the journey of digital transformation, the adoption of new technologies and methodologies is not just an operational upgrade but a strategic necessity. Leaders in the digital age are tasked with a pivotal role: to not only keep abreast of technological advancements but also to champion their adoption in ways that propel their teams and organizations forward. This commitment to innovation requires a discerning approach to embracing new tools and processes that enhance RWD practices and foster an environment of continuous learning and improvement.

The digital landscape is in a state of constant flux, with new technologies emerging at a rapid pace. Leaders must develop a keen eye for identifying which of these technologies have the potential to significantly impact their team's work. This involves staying informed through industry research, attending conferences, and engaging with professional networks. However, the adoption of new technology should always be strategic; leaders must evaluate the potential benefits against the team's current capabilities and the organization's overall goals. This discerning approach ensures that investments in technology drive meaningful improvements in efficiency, quality, and innovation.

Agile methodologies have transformed the way digital projects are managed, emphasizing flexibility, customer feedback, and iterative development. For leaders guiding RWD projects, adopting Agile practices means fostering a culture where change is not only expected but embraced as an opportunity for improvement. Agile principles encourage cross-functional collaboration, enabling designers, developers, and stakeholders to work closely together. This collaborative environment accelerates the design and development process, ensuring that the final product is closely aligned with user needs and business objectives.

Innovation is not without risk, but the fear of failure should never be a deterrent to trying new approaches. Leaders must promote a culture of experimentation, where team members feel empowered to explore new ideas and technologies. This environment encourages creativity and can lead to breakthrough solutions that set the organization apart from its competitors. Celebrating both successes and failures as learning opportunities reinforces the value of experimentation and keeps the team pushing the boundaries of what's possible.

The adoption of new technologies and methodologies often requires an investment in training and development to ensure that team members have the skills needed to leverage these tools effectively. Leaders should prioritize continuous learning, providing access to courses, workshops, and resources that help team members stay current with industry trends. This commitment to professional development not only enhances the team's capabilities but also boosts morale and job satisfaction, contributing to higher retention rates.

Ultimately, the successful integration of new technologies and methodologies into RWD practices hinges on leadership. Leaders must not only advocate for the adoption of innovative tools and processes but also lead by example. Demonstrating a willingness to learn, adapt, and innovate signals to the team that embracing change is both expected and rewarded. This leadership approach ensures that the team remains agile, responsive, and equipped to navigate the challenges of the digital landscape.

In conclusion, embracing new technologies and methodologies is a critical component of leadership in the age of digital transformation. By strategically integrating these tools into RWD practices, fostering a culture of experimentation and continuous learning, and leading by example, leaders can guide their teams to new heights of innovation and success.

Cultivating a culture of continuous learning

In the swiftly evolving terrain of digital transformation, the sustenance and advancement of design teams hinge significantly on cultivating a culture of continuous learning. This approach is not merely about staying abreast with the latest technological trends or mastering the newest design tools; it's about nurturing an environment where curiosity is encouraged, knowledge is shared freely, and professional growth is intertwined with the team's success. Leaders play a crucial role in fostering this culture, making it an indispensable part of the team's ethos.

In the digital age, learning cannot be viewed as a finite journey that concludes with formal education or a specific certification. Instead, it's a continuous path of growth and evolution. Leaders must emphasize the importance of lifelong learning, highlighting its relevance not just to individual career advancement but to the collective capability of the team to innovate and excel in RWD projects. By prioritizing learning as a core team value, leaders can inspire their members to seek out new knowledge and skills actively.

A culture of continuous learning requires tangible support mechanisms. Leaders can facilitate this by providing diverse opportunities for professional development. This might include sponsoring attendance at industry conferences, offering access to online courses and workshops, or organizing internal training sessions led by team members or external experts. Importantly, learning opportunities should not be confined to technical skills alone but should also encompass soft skills such as communication, leadership, and strategic thinking, which are equally vital in navigating the digital landscape.

One of the most effective ways to embed continuous learning into a team's culture is by encouraging knowledge sharing. Leaders can establish regular forums such as lunch-and-learns, tech talks, or show-and-tell sessions where team members present on recent projects, explore new tools and techniques, or discuss challenges and solutions. This not only helps disseminate knowledge across the team but also fosters a sense of community and collaboration.

A culture of continuous learning is inherently linked to experimentation and innovation. Leaders should create an environment where team members feel safe to experiment with new ideas without fear of failure. Encouraging risk-taking and viewing setbacks as valuable learning experiences can lead to breakthrough innovations and significant advancements in RWD practices. Celebrating both successes and learnings from failures reinforces the message that experimentation is a critical path to discovery and improvement.

Perhaps the most impactful way leaders can cultivate a culture of continuous learning is by leading by example. When leaders actively engage in learning new skills, share their insights, and demonstrate a genuine commitment to professional growth, it sets a powerful precedent for the team. This leadership approach not only inspires team members but also establishes a shared vision of perpetual growth and adaptation.

In conclusion, cultivating a culture of continuous learning within design teams is a critical leadership responsibility in the age of digital transformation. By emphasizing lifelong learning, creating opportunities for professional development, encouraging knowledge sharing, supporting experimentation, and leading by example, leaders can ensure their teams remain dynamic, innovative, and well equipped to meet the challenges of the digital future.

Leading by example in the digital era

Leading by example is a timeless principle of effective leadership, and in the digital era, its importance is magnified. As digital transformation reshapes industries, the need for leaders to model the behaviors and skills required to navigate this new landscape becomes paramount. For those steering design teams, especially within the realm of RWD, leading by example is not just about adopting new tools or techniques; it's about embodying the mindset and values that will drive success in a digital-first world.

In the digital era, leaders must demonstrate a willingness to embrace and learn new technologies. This doesn't mean leaders must become technical experts in every new tool or platform. However, having a solid understanding of the technologies that underpin digital projects — be it RWD principles, **content management systems** (**CMSs**), or analytics tools — enables leaders to make informed decisions and provide meaningful guidance. When leaders invest time in understanding the digital tools and platforms their teams use, it signals a commitment to the digital transformation journey.

Leadership in the digital era entails being an advocate for digital best practices. This involves championing the importance of UX, accessibility, and mobile-first design principles. By prioritizing these elements in projects and discussions, leaders set a standard for the team's work. It's about making clear that the goal isn't just to complete tasks but to create digital solutions that are intuitive, accessible, and effective across all devices and platforms.

The pace of change in the digital landscape requires teams to be innovative and agile. Leaders can exemplify this by promoting a culture where experimentation is encouraged and failure is seen as a learning opportunity. This means being open to trying new approaches, iterating on designs based on feedback, and pivoting when necessary. When leaders themselves are willing to question assumptions, test new ideas, and learn from outcomes, they inspire their teams to adopt the same growth-focused mindset.

In the digital era, learning is an ongoing process. Leaders can lead by example by actively pursuing their own professional development and sharing their learnings with the team. Whether it's through attending workshops, reading industry publications, or participating in online forums, leaders who are visibly committed to keeping their skills and knowledge up to date encourage their team members to do the same. This creates an environment where continuous improvement is the norm and staying ahead of digital trends is a shared responsibility.

Digital transformation often requires cross-functional collaboration and clear communication. Leaders can model effective communication by being transparent about goals, challenges, and successes. Utilizing digital tools to enhance team collaboration and ensuring that remote or distributed team members feel included and valued exemplifies how digital tools can be leveraged to build a cohesive team culture.

Leading by example in the digital era is about more than just adopting the latest technologies; it's about embodying the qualities that will enable teams to thrive amid digital transformation. Leaders who embrace technological proficiency, advocate for best practices, foster innovation, commit to continuous learning, and prioritize effective communication set a powerful example for their teams. This leadership approach not only drives digital projects forward but also cultivates a team culture that is resilient, adaptive, and poised for success in the digital landscape.

Let's explore the fundamental principles of RWD.

Effective communication of RWD principles

Understanding the foundations of RWD is crucial. This section is designed to provide leaders with a solid base to effectively communicate the core principles and benefits of RWD to their teams, ensuring a unified approach to designing and implementing responsive solutions.

RWD is built around three fundamental principles:

- **Fluid grids**: Websites are designed in a way that allows layout grids to flexibly adapt to any screen size. This fluidity ensures that content naturally fits across different devices, enhancing readability and usability without the need for device-specific designs.

- **Flexible images and media**: Images, videos, and other media formats are set to scale within their containing elements. This flexibility prevents visual content from exceeding the bounds of their containers, which is essential for maintaining a cohesive appearance on screens of varying sizes.

- **Media queries: Cascading Style Sheets (CSS)** technology is used to apply different style rules based on the device's characteristics, such as its width, resolution, and orientation. Media queries enable the website to present an optimized layout that corresponds to the user's device, improving the overall UX.

Effective leaders articulate the business value of adopting RWD principles, emphasizing benefits such as the following:

- **Improved UX:** A responsive site offers a seamless experience across devices, which can significantly reduce bounce rates and increase engagement.

- **Enhanced search engine optimization (SEO):** Google favors mobile-friendly websites. Adopting RWD can improve search rankings, driving more organic traffic.

- **Cost efficiency:** Maintaining a single responsive website is more cost-effective than operating separate sites for mobile and desktop users.

- **Future scalability:** RWD makes it easier to adapt to future devices and screen sizes, ensuring the website remains functional and appealing.

When communicating the foundations of RWD to stakeholders, focus on simplifying complex concepts into relatable benefits. Use real-world examples and case studies to illustrate how RWD principles have been successfully applied and the positive outcomes achieved. Encourage questions and discussions to ensure a deep understanding and buy-in.

Leaders should integrate RWD principles into project workflows through the following:

- **Training and development:** Providing team members with resources and training to understand and implement RWD effectively

- **Collaboration tools:** Utilizing design and prototyping tools that support responsive principles, facilitating collaboration across design and development teams

- **Quality assurance:** Implementing testing protocols that ensure designs perform as intended across various devices and browsers

Finally, leaders must lead by example by prioritizing responsive design in their projects and decision-making processes. Showcasing a commitment to RWD principles in every aspect of the project life cycle reinforces their importance and encourages the team to follow suit.

The foundations of RWD form the bedrock upon which effective communication and implementation strategies are built. Leaders play a pivotal role in disseminating these principles, ensuring their teams are well equipped to deliver responsive, user-centered solutions. By understanding and advocating for the core tenets of RWD, leaders can guide their teams toward creating digital experiences that are accessible, engaging, and future-proof.

There may be scenarios where we need to convince others of the benefits of RWD, which is why articulating the business value of RWD is so important.

Articulating the business value of RWD

Articulating the business value of RWD is a crucial skill for leaders guiding design teams. As digital landscapes evolve and consumer behaviors shift toward mobile-first engagements, the imperative for RWD becomes not just a design preference but a strategic business decision. Leaders must effectively communicate this value to stakeholders, team members, and clients to foster buy-in and prioritize RWD initiatives. This communication emphasizes how RWD aligns with broader business goals, such as enhancing UX, improving search engine rankings, and ultimately driving conversions and revenue.

The starting point in articulating the business value of RWD is framing it as a strategic asset rather than a technical necessity. Leaders should highlight how RWD extends the reach of a website to a broader audience, ensuring an optimal viewing experience across a wide range of devices. This adaptability is crucial in today's fragmented digital environment, where users access content via smartphones, tablets, laptops, and other emerging technologies.

A core aspect of the business value of RWD lies in its impact on UX and engagement. RWD ensures that users have a seamless and consistent experience, irrespective of the device they use, leading to higher satisfaction levels. Leaders should communicate how improved UX results in longer site visits, lower bounce rates, and increased engagement, which are critical metrics that influence a business's online success and brand perception.

Search engines, notably Google, have increasingly prioritized mobile-friendly websites in their ranking algorithms. Articulating the SEO benefits of RWD involves explaining how mobile optimization directly influences a site's visibility in search results. A responsive site is more likely to rank higher, especially in mobile searches, leading to increased organic traffic. Leaders should underscore the synergy between RWD and SEO strategies as a compelling business advantage, emphasizing the role of RWD in driving targeted traffic and improving online visibility.

Ultimately, businesses invest in digital platforms to drive conversions and revenue. Leaders must articulate how RWD directly contributes to these bottom-line objectives. By enhancing the UX, ensuring consistency across devices, and improving search engine rankings, RWD creates a conducive environment for conversions. Highlighting case studies or metrics that demonstrate improvements in conversion rates and revenue post-RWD implementation can be particularly persuasive.

Another key business value of RWD is its role in simplifying website maintenance and ensuring scalability. A responsive website eliminates the need for multiple versions of a site, reducing development and maintenance costs. Leaders should communicate how RWD, by its nature, makes a website more adaptable to future technological changes or user needs, protecting the business's digital investment.

Articulating the business value of RWD is about connecting the dots between technical solutions and strategic business outcomes. Leaders who effectively communicate how RWD enhances UX, improves SEO, drives conversions, and ensures long-term sustainability can garner the necessary support for RWD initiatives. By framing RWD as a critical component of a business's digital strategy, leaders can secure the commitment and resources needed to implement responsive design principles effectively, ensuring the organization remains competitive in a rapidly evolving digital marketplace.

Strategies for effective RWD education and training

In the rapidly evolving digital landscape, ensuring that design teams are well versed in RWD principles is crucial. As leaders, one of our pivotal roles is to facilitate effective RWD education and training, equipping our teams with the knowledge and skills they need to succeed. This endeavor involves more than just occasional workshops; it requires a strategic approach to learning that is continuous, comprehensive, and collaborative. Here are several strategies leaders can employ to foster an environment where RWD education and training thrive:

- Creating a structured training program tailored to the team's specific needs and skill levels is foundational. This program should cover all aspects of RWD, from basic concepts and coding techniques to advanced strategies for optimization and UX. Incorporating a mix of learning formats, such as online courses, in-person workshops, and hands-on projects, can cater to different learning preferences and reinforce the material through practical application.

- Bringing in external experts for specialized training sessions can provide fresh perspectives and deep insights into RWD trends and best practices. These experts could be industry leaders, authors of widely respected design methodologies, or practitioners known for innovative work in RWD. Their external viewpoints can invigorate the team, introduce new ideas, and highlight real-world applications of RWD principles.

- **Peer-to-peer** (**P2P**) learning is an invaluable tool for fostering a culture of continuous improvement and knowledge sharing. Encourage team members to lead informal training sessions on topics where they have expertise or have learned something new. This not only helps disseminate knowledge across the team but also boosts confidence and reinforces a collaborative team environment.

- Practical, hands-on experience is often the most effective way to learn. Design and implement projects specifically aimed at challenging the team to use RWD principles in new and innovative ways. These projects could be internal initiatives or pro bono work for non-profits, offering a real-world context to apply learning while contributing to a good cause.

- The internet is a treasure trove of resources for learning RWD. Encourage your team to engage with online communities, forums, and social media groups focused on web design and development, as well as in-person meetups in their local areas. Platforms such as GitHub, Stack Overflow, and specific design and development subreddits can be excellent sources of knowledge, offering insights into common challenges, solutions, and best practices.

- Integrate continuous feedback into the training process. After completing training modules or projects, gather feedback on what was learned, what could be improved, and how the knowledge is being applied in practice. This feedback loop can help refine the training program over time, ensuring it remains relevant and effective.

- Recognizing and rewarding progress is crucial in keeping the team motivated. Celebrate milestones and achievements in RWD training and project work. Whether through formal recognition, small rewards, or simply verbal acknowledgment, showing appreciation for the effort and progress made in mastering RWD can be a powerful motivator.

Effective RWD education and training require a strategic, multifaceted approach. By developing structured training programs, leveraging expert knowledge, encouraging peer learning, and fostering a culture of continuous improvement, leaders can ensure their teams are proficient in RWD. This not only enhances the team's capabilities but also positions the organization to succeed in an increasingly responsive digital world.

Collaborative approaches to RWD implementation

Collaborative approaches to RWD implementation are essential for creating cohesive and effective digital products. In today's interdisciplinary teams, where designers, developers, content strategists, and project managers work closely together, fostering a culture of collaboration is pivotal. As leaders, our role is to facilitate these collaborative efforts, ensuring that all team members are aligned with the project's goals, understand their roles in implementing RWD, and can effectively communicate across disciplines. Here are several strategies to encourage a collaborative approach to RWD implementation:

- Effective collaboration begins with clear communication. Establishing dedicated channels for project communication — such as project management tools, chat applications, and regular meetingsSPACE — ensures that information flows freely among team members. Clear communication helps in aligning everyone's understanding of the project requirements and the specific RWD principles being applied.

- A collaborative approach requires team members to have a basic understanding of each other's roles and how they intersect with RWD. Encourage cross-disciplinary learning sessions where designers can learn about coding basics and developers can gain insights into design principles. This mutual understanding bridges gaps between disciplines, enabling team members to anticipate needs, identify potential challenges early, and devise solutions that integrate seamlessly across the project.

- Pair programming, where two developers work together at one workstation, is a well-known Agile technique that can be adapted for RWD projects. Similarly, design pairing — pairing a designer with a developer — can foster collaboration from the initial stages of a project. These practices encourage real-time problem-solving and knowledge sharing and can lead to more innovative and effective RWD solutions.

- Design systems — a comprehensive set of design standards, documentation, and components — serve as a common language for teams implementing RWD. They streamline the design-to-development process, ensuring consistency across the project while allowing for the flexibility needed in responsive design. Encouraging the use and ongoing development of a design system promotes collaboration by providing clear guidelines and shared resources.

- Collaborative reviews, where team members from different disciplines come together to assess work, provide feedback, and make decisions, are crucial for RWD projects. These sessions should focus not only on critiquing the work but also on understanding the rationale behind design and development choices. Constructive feedback loops enhance the project's quality and foster a sense of shared ownership among team members.

- Recognizing and celebrating the achievements of collaborative efforts reinforces the value of teamwork and encourages continued cooperation. Highlighting successful implementations of RWD that resulted from effective collaboration can serve as motivation for future projects and help to build a strong, cohesive team culture.

Adopting collaborative approaches to RWD implementation is not without its challenges. It requires commitment from all team members and a supportive environment fostered by leadership. By establishing clear communication, fostering cross-disciplinary understanding, utilizing pairings, leveraging design systems, conducting joint reviews, and celebrating successes, leaders can guide their teams toward more effective and harmonious RWD implementations. This not only leads to better project outcomes but also contributes to a more engaging and productive team dynamic.

Feedback loops and RWD evolution

In the dynamic field of RWD, establishing robust feedback loops is vital for the evolution and refinement of digital projects. These loops facilitate continuous communication between design teams and stakeholders, including end users, ensuring that RWD initiatives remain aligned with user needs and business goals. Effective leaders understand the importance of integrating feedback loops into the RWD process, recognizing that they are crucial mechanisms for fostering innovation, enhancing user satisfaction, and driving project success.

To create effective feedback loops, leaders must first establish clear channels through which feedback can be gathered. This can include user surveys, analytics, usability testing sessions, and direct feedback from customer support interactions. Social media platforms and community forums also provide rich sources of informal feedback. Equally important is internal feedback within the project team and from cross-functional stakeholders, ensuring all perspectives are considered in the evolution of the RWD project.

Once feedback channels are established, the next step is integrating this feedback into the design and development process. This integration involves the following:

- **Regular review sessions**: Scheduling periodic meetings where the team can review feedback and identify trends or recurring issues that need addressing.

- **Prioritization**: Not all feedback will be equally critical to the project's success. Leaders must guide their teams in prioritizing feedback based on factors such as impact on UX, technical feasibility, and alignment with project goals.

- **Actionable plans**: Transforming feedback into actionable items is crucial. This may involve adjusting design elements, refining content for better clarity on various devices, or enhancing interactive features to improve usability.

Cultivating a culture that values responsiveness to feedback is essential. Teams should be encouraged to view feedback not as criticism but as opportunities for growth and improvement. Celebrating changes that were driven by user feedback reinforces the value of being responsive and agile. Leaders play a key role in modeling this attitude, demonstrating openness to change and commitment to using feedback constructively.

To close the feedback loop, it's important to measure the impact of any changes made in response to feedback. This can involve setting specific metrics for success, such as improved engagement metrics, reduced bounce rates, or higher conversion rates. Monitoring these metrics following changes allows the team to evaluate the effectiveness of their responses and further refine their approach.

In the context of RWD, feedback loops are not just tools for incremental improvement but catalysts for evolution. They enable teams to stay ahead of changing user behaviors, technology trends, and market demands. By continuously integrating feedback, RWD projects can adapt and evolve, ensuring they meet and exceed user expectations.

Feedback loops are integral to the successful implementation and continuous improvement of RWD projects. By establishing clear feedback channels, integrating feedback into the design process, promoting a culture of responsiveness, and measuring the impact of changes, leaders can ensure their RWD projects remain dynamic, user-centered, and aligned with strategic objectives. In doing so, they foster an environment where RWD can truly evolve in response to the needs of users and the goals of the organization.

How can we go about measuring the success of our implementation as well as the improvements that we've made?

Measuring success and iterative improvements

In the dynamic process of RWD, establishing clear benchmarks is essential for measuring success and guiding iterative improvements. These benchmarks serve as navigational beacons, helping teams evaluate the effectiveness of their RWD efforts against predefined criteria. For leaders, setting these benchmarks involves a balanced understanding of technical performance, UX, and business objectives. Here's how leaders can effectively set benchmarks for RWD success, ensuring their teams have concrete goals to strive toward and clear metrics to gauge their progress.

The first step in setting benchmarks is to clearly define what success looks like for the RWD project. This involves aligning with broader business goals — whether it's enhancing user engagement, increasing mobile traffic, improving search engine rankings, or boosting conversion rates. Understanding these objectives allows leaders to tailor benchmarks that directly contribute to achieving strategic outcomes.

Technical benchmarks are quantifiable and provide objective measures of a website's performance. Key metrics might include the following:

- **Load time**: The speed at which pages load on various devices, with specific targets set for mobile, tablet, and desktop
- **Responsiveness**: The time it takes for the website to adapt to different screen sizes or orientations
- **Accessibility scores**: Measured using tools that evaluate the site's compliance with accessibility guidelines (for example, the **Web Content Accessibility Guidelines (WCAG)**)

Setting precise targets for these metrics gives the development team clear performance goals and helps identify areas requiring optimization.

UX benchmarks focus on the qualitative aspects of how users interact with the website across different devices. These might include the following:

- **User satisfaction**: Measured through surveys, feedback forms, or **net promoter scores (NPS)**
- **Navigation efficiency**: The ease with which users can find information or complete tasks
- **Cross-device consistency**: Ensuring users have a seamless experience regardless of the device used

These benchmarks require regular user testing and feedback collection to assess how well the RWD meets user needs and expectations.

Ultimately, the success of RWD initiatives is measured by their impact on business goals. Benchmarks in this category might include the following:

- **Conversion rates**: The percentage of visitors completing a desired action, segmented by device type
- **Traffic metrics**: The volume of visitors and page views from various devices, monitoring shifts in mobile versus desktop usage

- **SEO rankings**: The website's position in search engine results for key queries, particularly in mobile searches

By linking RWD efforts to tangible business outcomes, leaders can demonstrate the value of their projects and secure ongoing support for optimization efforts.

Once benchmarks are established, they serve as the basis for ongoing evaluation and iterative improvement. Regularly reviewing performance against these benchmarks allows teams to identify areas of success and opportunities for further enhancement. It also fosters a culture of continuous learning and adaptation, critical in the ever-evolving landscape of web design and development.

Setting benchmarks for RWD success is a strategic exercise that requires a comprehensive understanding of technical specifications, user expectations, and business objectives. By establishing clear, measurable targets across these dimensions, leaders can provide their teams with the direction needed to create responsive, effective, and engaging web experiences. More importantly, these benchmarks lay the groundwork for iterative improvements, ensuring that RWD efforts remain aligned with changing technologies, user behaviors, and business goals.

Let's explore some tools and techniques that can be used to monitor RWD performance.

Tools and techniques for monitoring RWD performance

In the pursuit of excellence in RWD, monitoring performance is indispensable. The digital landscape offers a plethora of tools and techniques designed to gauge the effectiveness of RWD projects, providing invaluable insights into areas of success and those requiring improvement. For leaders, understanding and utilizing these resources is key to ensuring that their teams can iterate and enhance their designs effectively. This exploration focuses on the tools and techniques integral to monitoring RWD performance, aiding leaders in guiding their teams toward optimal outcomes.

Web analytics platforms, such as Google Analytics, offer a comprehensive overview of website performance across devices. These platforms can track a wide range of metrics, including user engagement, bounce rates, and conversion rates, segmented by device type. By analyzing this data, teams can identify how well their RWD is performing in real-world scenarios, understanding where users encounter friction and where the design excels.

Tools specifically designed for testing RWD allow teams to see how their websites render on various devices and screen sizes. Services such as BrowserStack and Responsinator provide real-time previews across multiple device configurations, helping identify visual and functional discrepancies that may not be apparent during the development phase. These tools are crucial for ensuring that the RWD meets its intended goals across the entire spectrum of user devices.

The speed and performance of a website are critical to its success, especially in a mobile-first world. Tools such as Google's PageSpeed Insights and WebPageTest offer detailed analyses of load times and performance bottlenecks. They provide specific recommendations for optimization, such as image compression, code minification, and leveraging browser caching. Regular use of these tools helps teams prioritize performance enhancements, a key aspect of RWD success.

Ensuring that RWD projects are accessible to all users, including those with disabilities, is a fundamental aspect of their success. Tools such as the WAVE Web Accessibility Evaluation Tool and axe Accessibility Checker allow teams to assess their websites against accessibility standards (for example, WCAG). These evaluations highlight issues that could hinder accessibility, guiding teams in making necessary adjustments to promote inclusivity.

Direct feedback from users can offer actionable insights that quantitative tools may not capture. Implementing feedback mechanisms, such as surveys, user forums, and comment sections, enables teams to gather subjective experiences of the website's responsiveness. This qualitative data is instrumental in understanding the user's perspective, informing iterations that enhance user satisfaction and engagement.

A/B testing platforms, such as Optimizely or Google Optimize, allow teams to test different versions of a webpage to see which performs better in terms of user engagement, conversion rates, or other key metrics. By systematically testing variations in design, layout, or content across devices, teams can make data-driven decisions that improve the effectiveness of their RWD projects.

Monitoring the performance of RWD projects requires a multifaceted approach, leveraging both quantitative and qualitative tools and techniques. By utilizing web analytics, responsive design testing tools, performance testing services, accessibility evaluation tools, user feedback mechanisms, and A/B testing platforms, leaders can provide their teams with the insights needed to refine and enhance their responsive designs continually. These tools not only help in measuring success but also in identifying opportunities for iterative improvements, ensuring that RWD projects remain aligned with user needs and business objectives in the ever-evolving digital landscape.

Analyzing user feedback and behavior

In the iterative process of refining RWD, analyzing user feedback and behavior stands as a cornerstone for success. This approach not only unveils how users interact with a website across different devices but also sheds light on their needs, preferences, and pain points. For leaders guiding design teams, harnessing insights from user feedback and behavior is pivotal in driving targeted improvements that enhance the overall UX and effectiveness of RWD. Here's how leaders can strategically approach this analysis to ensure their teams can make informed, user-centric design decisions:

- User feedback can be collected through various channels, including surveys, feedback forms embedded within the website, social media interactions, and direct customer support inquiries. Encourage users to share their experiences, specifically regarding how the website performs on different devices. Open-ended questions can reveal nuanced insights about user needs and expectations, while rating scales provide quantifiable data on user satisfaction levels.

- Tools such as Google Analytics, Hotjar, or Crazy Egg offer a wealth of information on how users navigate and interact with a website. These tools can track a variety of metrics, such as page views, bounce rates, and conversion paths, segmented by device type. More advanced features such as heatmaps and session recordings give a visual representation of user interactions, highlighting areas that attract the most attention and parts of the website that may cause confusion or frustration.

- Usability testing involves observing real users as they interact with the website across different devices. This can be conducted remotely using specialized software or in person for more in-depth analysis. Pay attention to how easily users can complete tasks, such as finding information or making a purchase, and note any obstacles they encounter. This direct observation is invaluable in understanding the practical challenges users face and identifying opportunities for improvement.

- Customer support interactions can be a rich source of insights into user issues and challenges. Analyzing queries and complaints related to website usability can help identify common problems that users encounter, such as navigation difficulties or content accessibility issues. This data can inform priorities for design iterations, ensuring that solutions address real user needs.

- Monitoring social media platforms for mentions of the website can provide spontaneous feedback from users. Social media listening tools can help track conversations about the website's usability and performance across devices. This feedback is often candid and can reveal immediate user reactions to design changes or updates.

- Once user feedback and behavior data have been collected, the next step is to synthesize these insights into actionable design improvements. Leaders should facilitate cross-functional discussions involving designers, developers, and content strategists to interpret the data and brainstorm solutions. Prioritizing issues based on their impact on UX and business goals ensures that the team focuses on the most critical areas for improvement.

Analyzing user feedback and behavior is a dynamic process that requires ongoing attention and responsiveness. By systematically collecting and interpreting user insights, leaders can empower their design teams to make data-driven decisions that refine RWD projects. This user-centric approach not only elevates the quality of the web experience across devices but also aligns design efforts with the evolving needs and expectations of the user base, fostering a stronger connection between the website and its audience.

Many of us use A/B testing across the realms of software development, but it's also applicable when approaching RWD optimization.

The role of A/B testing in RWD optimization

The optimization of RWD is a continuous process, where A/B testing plays a pivotal role in identifying the most effective design solutions. This methodical approach to testing allows leaders and their design teams to make data-driven decisions that enhance UX across various devices. By comparing two versions of a web page (A and B), teams can determine which elements contribute to better performance metrics, such as increased engagement, higher conversion rates, or improved user satisfaction. Here's a deeper look into the role of A/B testing in RWD optimization and how it can be effectively implemented:

- Before initiating A/B testing, it's crucial to define clear, measurable objectives based on the goals of the RWD project. Whether aiming to improve the mobile conversion rate, decrease the bounce rate on tablets, or enhance the navigation experience across all devices, having specific targets ensures the test's focus aligns with broader business and UX goals.

- In RWD, elements that may impact user interaction differently across devices include layout structures, **call-to-action** (**CTA**) buttons, menu designs, image sizes, and content presentation. Selecting which elements to test requires a strategic approach, often starting with areas known to influence **key performance indicators** (**KPIs**) significantly. Prioritizing these elements based on analytics insights and user feedback can guide the testing process more effectively.

- Implementing A/B testing in the context of RWD involves creating two versions of a page (A and B) that differ in one key aspect, such as the placement of a CTA button or the layout of a navigation menu. Traffic is then split between these versions, allowing for the collection of performance data across different devices. Utilizing A/B testing tools and platforms can simplify this process, offering features such as audience segmentation, real-time results tracking, and statistical analysis to validate the outcomes.

- The analysis phase is critical in A/B testing, as it reveals how different design choices impact user behavior and conversion rates across devices. It's essential to go beyond surface-level metrics, delving into device-specific data to understand the nuances of user interactions on mobile, tablet, and desktop. This analysis should inform whether the changes positively affect the objectives set forth at the test's inception.

- The insights gained from A/B testing should drive iterative improvements in the RWD project. Successful elements from the test can be rolled out across the site, while learnings from less successful outcomes can guide future design decisions. This iterative process is key to refining the UX continually, ensuring that the design remains effective as user expectations and device usage patterns evolve.

- Beyond the immediate benefits of optimizing RWD, A/B testing fosters a culture of empirical decision-making and continuous learning within design teams. Encouraging a mindset where hypotheses are tested, results are analyzed, and decisions are based on data is invaluable in navigating the complexities of designing for a multi-device world.

The role of A/B testing in RWD optimization cannot be overstated. It provides a scientific method to evaluate design decisions, ensuring that every change contributes positively to UX and business objectives. By embracing A/B testing, leaders can guide their teams in creating responsive designs that not only look great across all devices but also perform exceptionally, meeting both user needs and business goals.

It's the responsibility of the leaders to foster a culture that looks to improve.

Creating a culture of continuous improvement

Creating a culture of continuous improvement is pivotal for design teams, particularly in the dynamic realm of RWD. This culture not only drives the team toward excellence but also ensures adaptability in the face of evolving technological landscapes and user expectations. Leaders play a critical role in nurturing this environment, where the quest for betterment is ingrained in the team's ethos. Here are actionable strategies for embedding a culture of continuous improvement:

- Promote an environment where experimentation is valued and failures are seen as opportunities for learning rather than setbacks. By removing the fear of failure, leaders can inspire creativity and innovation within their teams. Encourage team members to undertake small-scale experiments in RWD projects, document their findings, and share these learnings with the team. This approach not only fosters a growth mindset but also contributes to the collective knowledge base.

- Schedule periodic sessions dedicated to reviewing projects, processes, and outcomes. Use these sessions not just to celebrate successes but to dissect what didn't work and why. Encourage open and constructive discussions where every team member can offer insights and suggestions. This reflective practice helps identify areas for improvement and reinforces the principle that continuous improvement is a team effort.

- Define specific, measurable goals related to RWD performance, UX, and project outcomes. These goals should challenge the team but remain achievable. Breaking down larger objectives into smaller, incremental milestones can help maintain motivation and provide a clear path to enhancement. Celebrate these milestones to recognize progress and keep the team focused on continuous improvement.

- Create platforms and opportunities for team members to share knowledge, insights, and new techniques related to RWD. Whether through internal workshops, lunch-and-learn sessions, or digital forums, encourage the exchange of ideas and best practices. This ongoing knowledge sharing is crucial for keeping the team updated on the latest trends and methodologies in responsive design.

- Support the team's professional growth by providing access to training, conferences, and learning resources. Professional development plays a key role in continuous improvement by equipping the team with new skills and perspectives. Encourage team members to pursue learning opportunities and share their acquired knowledge with the team, thus creating a cycle of learning and improvement.

- Leverage analytics, user feedback, and performance metrics to guide improvement efforts. Data-driven insights allow teams to make informed decisions about where to focus their improvement efforts. Regularly review these metrics and adjust strategies accordingly to ensure that RWD projects continually evolve to meet user needs and business goals.

- Leaders should embody the principles of continuous improvement in their actions and decisions. By demonstrating a commitment to personal growth, openness to feedback, and a willingness to adapt and learn, leaders can inspire their teams to embrace the same values. Leading by example is perhaps the most powerful tool for cultivating a culture of continuous improvement.

Establishing a culture of continuous improvement is essential for design teams navigating the complexities of RWD. Through experimentation, reflection, goal setting, knowledge sharing, professional development, data-driven insights, and leading by example, leaders can foster an environment where continuous improvement is not just encouraged but becomes a natural part of the team's workflow. This culture not only elevates the quality of RWD projects but also prepares the team to adapt and thrive in an ever-changing digital landscape.

Let's explore how we can maximize the work environment through creativity and problem-solving.

Nurturing creativity and problem-solving

Fostering a creative environment within design teams, especially those focused on RWD, is crucial for innovation and problem-solving. Creative environments encourage team members to explore new ideas, challenge assumptions, and develop unique solutions that enhance the UX across various devices. For leaders, creating such an atmosphere involves more than just verbal encouragement; it requires strategic actions and a commitment to cultivating a workspace that values and nurtures creativity at every level. Here's how leaders can effectively foster a creative environment for their RWD teams:

- Creativity thrives in environments where experimentation is welcomed and risk-taking is not penalized. Leaders should encourage their teams to try new approaches in RWD projects, even if they deviate from conventional methods. By allowing team members to explore and experiment, leaders can unlock innovative solutions that could significantly improve project outcomes. It's important to establish a safety net for failure, ensuring team members feel supported and valued, even when experiments don't yield the expected results.

- Access to the right tools and resources can significantly enhance a team's creative capabilities. Leaders should ensure their teams have what they need to bring their ideas to life, whether it's software, hardware, or access to new technologies. Additionally, investing in ongoing training and professional development can inspire creativity by exposing team members to the latest trends, techniques, and best practices in RWD and broader web design principles.

- Collaboration is a key driver of creativity. Designing physical and virtual spaces that facilitate easy collaboration among team members can spark new ideas and innovative problem-solving approaches. This might include open workspaces that encourage casual interactions, dedicated brainstorming rooms, or digital platforms that support seamless collaboration for remote teams. The goal is to make it easy for team members to share ideas, provide feedback, and work together on creative solutions.

- Diverse teams bring a wide range of perspectives, experiences, and ideas, enriching the creative process. Leaders should strive to build teams with diverse backgrounds, skill sets, and ways of thinking. Fostering an inclusive environment where every team member feels valued and heard is essential for unlocking the full creative potential of the team. Inclusion ensures that different viewpoints are considered, leading to more comprehensive and innovative RWD solutions.

- Dedicating time to exploration and personal projects can significantly boost creativity. Leaders might allocate certain hours or days for team members to work on projects they're passionate about, even if they're not directly related to current RWD tasks. This not only recharges their creative energies but can also lead to unexpected discoveries and innovations that benefit the team's primary projects.

- Recognizing and celebrating creative achievements encourages a culture of innovation. Highlighting successful projects, innovative solutions, and even creative attempts that didn't pan out as expected can motivate the team to continue pushing creative boundaries. Celebrations can be formal, such as awards or recognitions, or informal, such as team shout-outs during meetings.

Fostering a creative environment for RWD teams requires intentional strategies and actions from leadership. By encouraging experimentation, providing the necessary tools and resources, creating spaces for collaboration, promoting diversity and inclusion, allowing time for exploration, and celebrating creativity, leaders can cultivate an atmosphere where innovation flourishes. Such an environment not only enhances the team's work on RWD projects but also contributes to a fulfilling and engaging workplace culture.

Encouraging collaborative problem-solving

Encouraging collaborative problem-solving within design teams, particularly those focused on RWD, is essential for creating innovative, user-centric solutions that work across various devices and platforms. Collaboration brings diverse perspectives and expertise to the table, enabling teams to tackle complex problems more effectively than individuals working in isolation. Here's how leaders can cultivate an environment that encourages collaborative problem-solving among their RWD teams:

- Effective collaboration starts with clear communication. Leaders should establish and maintain open lines of communication within the team, ensuring that every member feels comfortable sharing their ideas and feedback. This can involve regular team meetings, using collaborative tools for digital communication, and encouraging open discussions about ongoing projects. Clear communication helps prevent misunderstandings and ensures that all team members are on the same page.

- For collaborative problem-solving to be effective, a culture of mutual respect is essential. Team members should feel valued for their contributions and confident that their ideas will be considered respectfully. Leaders can foster this culture by modeling respect in their interactions and addressing any behaviors that undermine team cohesion. When team members respect each other, they are more likely to listen, contribute, and work together effectively.

- One of the strengths of collaborative problem-solving is the ability to leverage the diverse skill sets within a team. Leaders should encourage team members to bring their unique strengths to the table, whether it's in design, development, UX, or another area relevant to RWD. By recognizing and utilizing these diverse skills, teams can approach problems from multiple angles and develop more comprehensive solutions.

- Workshops and brainstorming sessions are effective ways to engage team members in collaborative problem-solving. Leaders can organize workshops focused on specific challenges or projects, inviting input from all team members. These sessions should be structured to encourage participation from everyone, using techniques such as brainstorming, design thinking, and prototype testing to explore solutions collaboratively.

- Collaborative problem-solving shouldn't be limited to the RWD team alone. Encouraging collaboration with other departments, such as marketing, sales, and customer service, can provide additional insights and perspectives that enhance problem-solving efforts. Leaders can facilitate cross-functional collaboration by organizing joint meetings, shared projects, and interdepartmental workshops that focus on shared goals and challenges.

- Recognizing and rewarding collaborative efforts can motivate teams to continue working together effectively. Leaders should highlight successful collaborative problem-solving instances, acknowledging the contributions of all team members involved. Rewards can be formal, such as bonuses or promotions, or informal, such as public recognition or team celebrations. This not only reinforces the value of collaboration but also builds a strong sense of team unity.

- Adopting tools and technologies that facilitate collaboration can significantly enhance the problem-solving process. This might include project management software, collaborative design tools, and platforms for real-time communication and document sharing. Leaders should ensure the team has access to these tools and is trained to use them effectively, streamlining collaboration and making it easier to work together on complex RWD projects.

Encouraging collaborative problem-solving within RWD teams is crucial for developing innovative, effective solutions. By establishing clear communication channels, promoting a culture of mutual respect, leveraging diverse skill sets, facilitating collaborative workshops, encouraging cross-functional collaboration, recognizing and rewarding collaboration, and implementing supportive tools and technologies, leaders can create an environment where collaborative problem-solving thrives. This approach not only leads to better RWD outcomes but also fosters a more engaged, cohesive, and productive team.

Let's revisit design thinking next.

Implementing design thinking

Implementing design thinking within RWD teams is a strategic approach that can significantly enhance creativity and problem-solving capabilities. Design thinking is a user-centered methodology that encourages teams to focus on the needs and experiences of users, fostering a deep understanding of the challenges they face. This process involves empathy, ideation, prototyping, testing, and iteration, enabling teams to develop innovative solutions that are both effective and user-friendly. Here's how leaders can guide their teams in implementing design thinking to improve their RWD projects:

- The first step in design thinking is to cultivate empathy for the users. Leaders should encourage their teams to conduct thorough user research, including interviews, surveys, and observation, to gain insights into the users' needs, preferences, and behaviors across different devices. Understanding the user's perspective is crucial for identifying the real problems that need solving. Teams should be reminded that empathy involves more than just collecting data; it's about connecting with users on a human level to understand their experiences and emotions.

- With a solid understanding of user needs, teams can move into the ideation phase. Leaders should facilitate brainstorming and ideation sessions, creating a safe space for free-flowing creativity and the generation of a broad range of ideas. Encourage the team to think outside the box and consider all possible solutions, no matter how unconventional. The goal of ideation is to explore a wide spectrum of options that can address user challenges identified during the empathy phase.

- Prototyping is a critical component of design thinking, allowing teams to bring their ideas to life. Leaders should ensure their teams have the resources and time to develop prototypes for their RWD solutions. These prototypes do not need to be polished or complete; they simply need to be functional enough to test the core concepts of the proposed solutions. Encouraging rapid prototyping helps teams iterate quickly, learn from their mistakes, and refine their ideas based on practical insights.

- Testing prototypes with real users is essential for validating the effectiveness of design solutions. Leaders should organize user testing sessions, guiding teams to observe how users interact with the prototypes on different devices. Collecting user feedback during these sessions is invaluable for understanding what works, what doesn't, and why. This direct input from users helps teams to iterate on their designs with a clear focus on improving UX.

- Design thinking is inherently iterative, recognizing that the first solution is rarely the best solution. Leaders should foster an environment where iteration is embraced as a natural and necessary part of the design process. Encourage the team to view each iteration as an opportunity to learn and improve rather than as a setback. This mindset ensures that the team remains agile, responsive to user feedback, and committed to refining their RWD solutions until they meet the users' needs effectively.

Implementing design thinking within RWD teams offers a structured yet flexible approach to creativity and problem-solving. By focusing on user needs, encouraging broad ideation, prototyping rapidly, testing with users, and embracing iteration, leaders can guide their teams in developing more innovative, user-centered solutions. This methodology not only enhances the team's problem-solving capabilities but also aligns closely with the goals of RWD, ensuring that digital experiences are accessible, engaging, and satisfying across all devices.

Now, let's explore how we can continuously improve through the implementation of Agile.

Adapting to challenges with Agile methodologies

Adapting to challenges with agile methodologies in the context of leading and guiding design teams, particularly those focused on RWD, is pivotal for nurturing creativity and solving problems efficiently. Agile methodologies, with their emphasis on flexibility, collaboration, and iterative development, provide a robust framework for teams to adapt to design challenges dynamically. This approach encourages rapid experimentation, continuous feedback, and the ability to pivot when necessary, aligning perfectly with the unpredictable and fast-paced nature of web design projects.

Agile methodologies prioritize adaptive planning and evolutionary development. For RWD teams, this means the freedom to adjust designs based on new information or changing user needs without being strictly bound to a rigid project plan. Leaders should encourage teams to remain open to changing requirements and to view such changes not as setbacks but as opportunities to refine and improve the design to better meet user expectations.

One of the core principles of Agile is cross-functional collaboration. In the context of RWD, this involves designers, developers, content strategists, and testers working closely together throughout the project life cycle. Leaders can facilitate this by organizing regular stand-up meetings, sprint planning sessions, and retrospectives to ensure all team members are aligned on goals, progress, and challenges. This collaboration fosters a deeper understanding of different aspects of the project among team members, enhancing creativity and problem-solving capabilities.

Agile methodologies emphasize the importance of delivering working software early and frequently. For RWD teams, this translates to the development of prototypes and **minimum viable products** (**MVPs**) that can be quickly tested and iterated upon. Leaders should encourage rapid prototyping as a means to explore creative solutions and validate ideas with real users. This iterative process allows the team to learn from each cycle, making data-driven decisions to refine the design progressively.

Continuous feedback is a cornerstone of Agile. RWD teams should be encouraged to seek ongoing feedback from users, stakeholders, and team members throughout the project. Leaders can facilitate this by setting up regular review sessions, user testing, and feedback channels that allow for quick responses to insights gathered. This approach ensures that the design remains user-centered and that any issues can be addressed promptly, keeping the project on track and aligned with user expectations.

Agile methodologies also advocate for reflective practice and continuous learning. After each iteration or project phase, leaders should guide their teams in conducting retrospectives to reflect on what worked well, what didn't, and why. This practice encourages the team to acknowledge successes, learn from mistakes, and identify areas for improvement. By fostering an environment where learning from every experience is valued, leaders can help their teams grow more adept at navigating challenges and innovating solutions.

Adapting to challenges with Agile methodologies offers RWD teams a flexible, collaborative, and iterative approach to design and problem-solving. By embracing these practices, leaders can nurture a creative, adaptive, and efficient team culture that is well equipped to meet the demands of modern web design projects. Agile's emphasis on user feedback, cross-disciplinary collaboration, and continuous improvement aligns closely with the goals of RWD, ensuring that teams can create responsive, user-friendly, and high-quality web experiences.

Measuring and encouraging creative outcomes

Measuring and encouraging creative outcomes within design teams, especially those focused on RWD, involves both acknowledging the inherently qualitative nature of creativity and applying quantitative metrics to gauge effectiveness and progress. Leaders have a critical role in fostering an environment that not only values innovation and original thinking but also understands how to recognize and cultivate these attributes. This balance ensures that creativity is not just a buzzword but a tangible, driving force behind the team's success.

Before leaders can measure creativity, it's essential to define what creative outcomes look like within the context of RWD. These outcomes often manifest as innovative solutions to design challenges, unique UX enhancements, or novel approaches to **user interface** (**UI**) design that improve responsiveness across devices. Leaders should work with their teams to establish clear criteria for what constitutes a "creative outcome," considering both the process (how ideas are generated and developed) and the product (the end result of these creative endeavors).

While creativity can be elusive to quantify, leaders can employ several methods to measure it indirectly. Metrics might include the number of new ideas generated in brainstorming sessions, the percentage of those ideas that progress to prototyping, or the impact of implemented creative solutions on user engagement and satisfaction. Other qualitative measures, such as peer reviews or feedback from stakeholders, can provide insights into the perceived novelty and usefulness of creative outputs.

Encouraging creative outcomes requires more than just measurement; it necessitates a supportive environment that champions risk-taking, values diverse perspectives, and allows for failure. Leaders should cultivate a culture where team members feel safe to express unconventional ideas without fear of criticism or failure. This involves providing time and resources for exploration and experimentation, as well as recognizing and celebrating creative achievements, regardless of their immediate practical application.

Diversity in team composition, including different backgrounds, skill sets, and ways of thinking, can significantly enhance creative output. Leaders should strive to assemble teams with a broad range of experiences and expertise, as this diversity fosters a rich breeding ground for innovative ideas. Encouraging collaboration across these varied perspectives can lead to more comprehensive and creative solutions to RWD challenges.

A commitment to ongoing learning and skill development is crucial for nurturing creativity. Leaders should encourage their teams to stay abreast of the latest design trends, technologies, and methodologies, providing opportunities for training, workshops, and attendance at industry conferences. This not only enhances the team's skill set but also inspires new ideas and approaches that can feed into creative processes.

Feedback loops are essential for refining creative ideas and ensuring they meet project goals and user needs. Leaders should establish regular check-ins, critiques, and user testing sessions to evaluate creative outcomes against defined objectives and metrics. Constructive feedback helps to refine ideas, improve designs, and ensure that creative efforts contribute meaningfully to project success.

Measuring and encouraging creative outcomes in RWD teams involves a combination of defining what creativity means in the context of web design, setting up appropriate metrics, and creating an environment that nurtures innovative thinking. By valuing diversity, promoting continuous learning, and implementing feedback loops, leaders can ensure that creativity is a pivotal and measurable component of their team's approach to responsive design, leading to more engaging, effective, and user-centered web experiences.

Summary

We've unpacked a multitude of strategies and insights pivotal for leadership in the ever-evolving domain of RWD. Reflecting on our exploration, we've covered essential ground, from developing leadership qualities that resonate with the demands of the digital age to mastering the nuanced art of communicating RWD principles effectively within diverse team settings. We delved into fostering an environment ripe for innovation, where creativity and problem-solving flourish, and we examined critical metrics that define success and iterative processes that ensure continual refinement and growth.

The skills and lessons imparted in this chapter are invaluable tools in the arsenal of anyone aspiring to lead design teams toward achieving exceptional outcomes. Understanding how to navigate digital transformation, embrace new technologies, and cultivate a culture of continuous learning are foundational to thriving in the fast-paced digital landscape. Effective communication of RWD principles ensures that teams are aligned, informed, and motivated while nurturing creativity and innovation leads to groundbreaking solutions that stand out in a crowded marketplace. Moreover, grasping the importance of measuring success and embracing a cycle of feedback and improvement is key to maintaining relevance and exceeding user expectations.

These competencies are not just about leading teams; they're about inspiring change, driving progress, and making a tangible impact on the digital products that shape our world. They equip leaders with the ability to not only adapt to change but to be the harbingers of it, fostering environments where creativity, efficiency, and excellence are the norms.

As we close this chapter, we stand at the threshold of the next leg of our journey — *Chapter 11*. Building on the foundations laid in this chapter, we'll delve into ensuring the quality of responsive designs and quantifying their **return on investment** (**ROI**). This natural progression will take us from the realms of creation and leadership into evaluation and optimization, exploring how to measure the efficacy of our efforts and ensure they deliver not only in terms of UX but also in tangible business value. The upcoming chapter promises to equip readers with the insights needed to not only design and lead effectively but also to assess and enhance the impact of their work in the broader context of business success and user satisfaction.

11

Quality Assurance and ROI

Quality Assurance (QA) stands at the forefront of ensuring the success of RWD projects, acting as the critical bridge between innovative design and the final user experience. This chapter delves into the indispensable role of QA in validating the functionality, usability, and performance of RWD across a myriad of devices and platforms. By embracing comprehensive testing strategies tailored to address the unique challenges of responsive designs, teams can significantly elevate the quality of their digital offerings, ensuring that they meet both user expectations and business objectives.

We will guide you through the intricacies of implementing robust QA processes that are specifically designed for the nuanced demands of RWD. From initial planning stages to final deployment, understanding how to effectively apply QA principles in this context is crucial for identifying and rectifying potential issues that could compromise the user experience, or hinder performance across devices.

Moreover, this chapter underscores the importance of accurately measuring the ROI of RWD initiatives. Through a blend of quantitative metrics and qualitative insights, we'll explore methodologies to evaluate the tangible and intangible benefits of responsive designs, providing a clear picture of their impact on business success. This analysis not only justifies the investment in RWD but also informs future strategies and improvements.

Lastly, we will address the challenge of maintaining a balance between rigorous QA practices and the constraints of project timelines. Efficiently integrating QA processes without compromising the project schedule demands, strategic planning, prioritization, and the adoption of agile methodologies. By highlighting best practices and effective tools, we aim to equip you with the knowledge to navigate this balancing act, ensuring timely project completion without sacrificing the quality of the end product.

As we journey through the critical facets of QA in the context of RWD, our goal is to arm you with the skills and insights necessary to champion quality, drive user satisfaction, and achieve remarkable ROI from your responsive design projects.

The topics covered in this chapter are as follows:

- The critical role of testing in RWD

- Aligning QA goals with business outcomes

- Continuous feedback and iteration

- Measuring the tangible ROI of responsive initiatives

The critical role of testing in RWD

In the dynamic realm of RWD, understanding the spectrum of testing is paramount to ensure that websites deliver optimal user experiences across a multitude of devices. RWD testing is not a one-size-fits-all process; it encompasses a variety of tests, each designed to address specific aspects of a website's functionality, appearance, and performance on different screen sizes and resolutions. This comprehensive approach to testing is crucial for identifying and resolving issues that could detract from the user experience or hinder the site's performance.

Types of RWD testing

Testing RWD can be broken down into a couple of different approaches:

- **Visual testing**: This involves checking the visual aspects of a website across devices to ensure that layouts adjust appropriately, images scale correctly, and typography remains readable without manual adjustments by the user. Visual testing helps maintain design integrity and consistency, which are crucial for brand identity and user trust.

- **Functionality testing**: Beyond looking right, responsive sites must also function correctly on various devices. Functionality testing verifies that all interactive elements, such as buttons, links, and forms, work as intended, regardless of the device. This testing phase is essential for ensuring a seamless user journey across platforms.

- **Performance testing**: Speed and responsiveness are key components of user satisfaction. Performance testing evaluates how quickly pages load and how smoothly they operate on different devices and network conditions. This type of testing aims to minimize load times and enhance the overall responsiveness of a website, contributing to a positive user experience.

- **Usability testing**: This user-centered testing method focuses on how real users interact with a website on different devices. Usability testing can uncover practical challenges users may face, such as navigation difficulties or content accessibility issues, providing valuable insights to improve the site's design and functionality.

- **Cross-browser testing**: Websites must perform well across various browsers, not just different devices. Cross-browser testing ensures that a site's features and functionalities are consistent and reliable on all major browsers, addressing any compatibility issues that could affect user experience. It should be noted that this has become a bit less important in the last few years, due to browser vendors better adhering to standards.

These are the five ways that RWD can be tested.

Implementing RWD testing

Successful implementation of RWD testing requires a strategic approach that incorporates various techniques to achieve a thorough testing process:

- **Automated testing tools**: Leveraging automated testing tools can significantly streamline the testing process, allowing for the efficient identification of layout issues, broken links, and performance bottlenecks across a wide range of devices and browsers. Some examples include Cypress or Selenium.

- **Real device testing**: While simulators and emulators offer a preliminary glimpse into how a website might perform, testing on real devices provides the most accurate insights into the user experience. Real device testing helps uncover issues that may not be apparent through emulation, such as touch responsiveness and real-world performance.

- **Continuous integration**: Integrating testing into the continuous development process allows for the early detection and resolution of issues. This approach supports a more agile development cycle, where testing and feedback loops inform ongoing design and development efforts.

Understanding the spectrum of RWD testing and implementing a comprehensive testing strategy are crucial for delivering websites that not only look great but also perform flawlessly across all devices. This multifaceted approach to testing ensures that responsive sites meet the high standards expected by users and stakeholders alike, ultimately contributing to a site's success and ROI.

Implementing effective cross-device testing strategies

Implementing effective cross-device testing strategies is crucial in the realm of RWD, where ensuring a seamless user experience across a wide array of devices is paramount. As the diversity of screen sizes, resolutions, and operating systems continues to expand, RWD projects demand a comprehensive approach to testing that can adapt to the complexities of the modern digital landscape. This page explores key strategies to conduct thorough cross-device testing, ensuring that responsive websites perform optimally for every user, regardless of their choice of device.

Let's look at the strategies:

- The first step in an effective cross-device testing strategy is to prioritize which devices and browsers to test based on real user data. Utilizing web analytics can reveal which devices and browsers your target audience uses most frequently, allowing you to focus your testing efforts on where they will have the greatest impact. This data-driven approach ensures that resources are allocated efficiently, covering the most critical user segments first.

While testing on real devices provides the most accurate insights into user experience, it's not always feasible to test on every device in the market. Combining real device testing with the use of emulators and simulators can offer a balanced approach. Emulators simulate the software and hardware environments of different devices, while simulators mimic the behavior of the software. This combination allows for broad coverage and can help identify device-specific issues early in the development process.

- Several tools are specifically designed to facilitate cross-device testing for RWD projects. These tools can automatically adjust browser window sizes to simulate various devices, test responsive breakpoints, and identify layout issues. Utilizing these tools can significantly streamline the testing process, allowing for the quick identification and resolution of design and functionality issues across device types.

- Visual regression testing involves capturing screenshots of web pages across different devices and comparing them to identify unintended changes or issues. This type of testing is invaluable for ensuring visual consistency and detecting layout problems that might not be apparent through code-based testing alone. Automated tools can help manage the vast number of comparisons needed for thorough cross-device visual testing.

- Beyond visual consistency, it's essential to test how interactive elements of a website behave on various devices. This includes testing touch interactions on touchscreens, hover effects on desktops, and the overall responsiveness of the website's interactive features. Interaction testing ensures that the website not only looks right but also functions correctly, providing a smooth and intuitive user experience.

- Cloud-based testing platforms offer access to a wide range of devices, browsers, and operating systems, facilitating comprehensive cross-device testing without the need for a large inventory of physical devices. These platforms can provide instant access to the latest devices and browser versions, making it easier to test under real-world conditions and keep up with the rapidly evolving device market.

Effective cross-device testing is not a one-time task but a continuous part of the RWD process. Encouraging a culture of ongoing testing and iteration within a team ensures that new content, features, and design changes are consistently evaluated across devices. This ongoing commitment to QA helps maintain a high standard of user experience, regardless of how or where a website is accessed.

Implementing these strategies for cross-device testing is fundamental to the success of any RWD project. By prioritizing testing efforts based on user data, utilizing a mix of real devices and emulators, leveraging specialized tools, and fostering a culture of continuous testing, teams can ensure that their responsive designs meet the diverse needs of today's web users, ultimately enhancing the quality and reach of their digital presence.

Usability testing for an optimal user experience

As digital platforms diversify and user access points multiply, it's imperative for design and development teams to adopt a robust approach to testing. This ensures that web solutions not only look aesthetically pleasing across devices but also function flawlessly, offering a seamless user experience. Here's a strategic approach to implementing effective cross-device testing for RWD projects to optimize the user experience:

- Develop a comprehensive device matrix that lists the devices, operating systems, and browsers that your target audience uses. This matrix should be informed by analytics data to prioritize testing on platforms where your users are most active. It's essential to include a variety of screen sizes, resolutions, and input methods (touch, mouse, and stylus) to cover the spectrum of user interactions.

- Leverage tools specifically designed for responsive testing. These can range from browser developer tools, allowing you to simulate different screen sizes, to more sophisticated software, enabling real-time testing across multiple devices and platforms simultaneously. Tools such as BrowserStack, Sauce Labs, and CrossBrowserTesting offer cloud-based testing environments that can significantly expand your testing capabilities without the need for an extensive physical device library.

- Automate repetitive testing tasks to save time and ensure consistency in testing. Automation tools can help you perform regression testing every time a change is made, ensuring that new code doesn't break existing functionality across different devices. Frameworks such as Selenium or Appium can automate web browser testing, while visual regression tools can automatically detect visual discrepancies across device layouts.

- While emulators and simulators are valuable for early-stage testing and debugging, nothing beats testing on real devices. Real device testing helps you understand exactly how your website performs under real-world conditions, including factors such as touch responsiveness, mobile network latency, and device-specific quirks. Incorporating a mix of popular and niche devices, especially those commonly used by your target audience, will provide a well-rounded view of your website's performance.

- Automated tests are efficient for covering ground quickly, but manual exploratory testing is invaluable for uncovering usability issues that automated tests might miss. Encourage testers to use the site as real users would, exploring different paths, interactions, and features. This type of testing is particularly effective for identifying problems with navigation, content accessibility, and the overall user experience on different devices.

- Ensure your RWD project is accessible to all users, including those with disabilities. Use both automated tools and manual testing to check compliance with web accessibility standards, such as the **Web Content Accessibility Guidelines** (**WCAG**). Testing with screen readers, keyboard navigation, and other assistive technologies is critical to ensure that your responsive site is usable by everyone, regardless of device or ability.

- Create a feedback loop that includes developers, designers, QA testers, and, if possible, real users. Continuous feedback helps to quickly identify and address issues that arise during the development process. Encouraging an environment where feedback is actively sought and acted upon ensures that the final product is not only technically sound but also meets user expectations and business goals.

Implementing these cross-device testing strategies requires a proactive, organized approach, but it pays dividends in creating RWD solutions that stand up to the demands of today's diverse and dynamic digital landscape. By ensuring thorough testing across devices, teams can confidently deliver websites that offer exceptional user experiences, driving satisfaction, engagement, and ultimately, ROI.

The next step is to explore how to best connect QA with business.

Aligning QA goals with business outcomes

Aligning QA goals with business outcomes for RWD projects is essential for ensuring that technical efforts contribute directly to the achievement of broader organizational objectives. One of the most effective ways to establish this alignment is through the identification and monitoring of **Key Performance Indicators** (**KPIs**). KPIs serve as quantifiable metrics that reflect the success or progress of a project in terms of meeting its specified goals. For RWD projects, these indicators not only gauge the technical performance and user experience of a website across different devices but also how these factors impact business success.

The first step in leveraging KPIs effectively is to define which metrics are most relevant to both the QA aspects of RWD and the overarching business objectives. These KPIs can vary significantly, depending on the specific goals of the project and the nature of the business. Commonly, they include the following:

- **Page load time**: A critical factor in user experience, particularly on mobile devices. Faster load times are associated with higher user satisfaction and engagement rates.

- **Bounce rate**: The percentage of visitors who navigate away from the site after viewing only one page. A lower bounce rate indicates that users find the site engaging and are motivated to explore more content.

- **Conversion rate**: For e-commerce sites or pages with specific calls to action, the conversion rate measures the percentage of visitors who take the desired action. This is directly tied to revenue and is a clear indicator of business success.

- **Cross-device usability**: Metrics that assess the effectiveness of the site's RWD across different devices, such as device-specific conversion rates or engagement metrics.

- **SEO ranking**: Search engine rankings for key terms, which can influence traffic volume and quality.

Once relevant KPIs have been identified, the next step is to align them with specific business goals. This alignment ensures that QA efforts in improving these metrics will directly contribute to achieving business outcomes, such as increased sales, higher customer retention, or expanded market reach. For instance, if a business goal is to increase online sales through mobile traffic, focusing on KPIs related to mobile page load time, mobile conversion rates, and mobile user engagement becomes critical.

With KPIs defined and aligned with business goals, continuous monitoring and analysis are vital. This involves setting up tools and processes to track these metrics in real time and analyze them for insights. Tools such as Google Analytics, heat mapping software, and custom dashboards can provide ongoing visibility into KPI performance. Regular analysis helps identify trends, pinpoint issues, and uncover opportunities for optimization.

The insights gained from monitoring KPIs should feed directly back into the RWD project in a continuous loop of improvement. If certain metrics underperform, this can indicate areas where the responsive design needs refinement, whether it's optimizing images for faster loading, adjusting a layout for better usability on mobile devices, or enhancing content for SEO. By tightly integrating this feedback into iterative design and development processes, teams can ensure that their RWD projects remain aligned with both user needs and business objectives.

Identifying and focusing on the right KPIs for RWD projects enables businesses to bridge the gap between technical QA efforts and tangible business outcomes. This strategic approach ensures that every aspect of responsive design — from speed and usability to content and SEO — is optimized not just for performance across devices but also to contribute to an organization's success.

Integrating user feedback into QA processes

Integrating user feedback into QA processes is a transformative strategy to align the goals of RWD projects with tangible business outcomes. User feedback serves as a direct line to understanding how real-world users interact with a website across various devices, providing invaluable insights that can guide the refinement of RWD initiatives. This user-centric approach ensures that QA efforts are not just about ticking off technical compliance boxes but are also deeply connected to enhancing user satisfaction and driving business success.

The integration of user feedback into the QA process involves several key steps, each designed to capture, analyze, and act upon the insights provided by users. This process begins with establishing effective channels to gather feedback, which can include the following:

- **Surveys and questionnaires**: Deployed directly on the website or sent via email, these tools can gather targeted insights on user experiences and perceptions

- **Usability testing**: Conducting formal usability tests with participants representing the target audience can uncover usability issues that might not be apparent to designers and developers

- **Analytics and behavior tracking**: Analyzing user behavior through tools such as Google Analytics or heat mapping software can indirectly capture feedback by highlighting areas of a site where users experience difficulties

- **Social media and customer support**: Feedback gathered from social media channels and customer support interactions can also provide candid insights into user experiences and issues

Once feedback is collected, the next step is to analyze the data to identify common themes, recurring issues, and areas for improvement. This analysis should focus on issues that impact user experience across different devices, such as navigation problems, content readability, or interaction difficulties. Prioritizing these insights based on their potential impact on user satisfaction and business goals is crucial for effective resource allocation.

Integrating user feedback into the QA and development cycles involves several key actions:

- **Prioritization**: Based on the analysis, prioritize feedback that aligns with critical business outcomes, such as improving conversion rates or reducing bounce rates.

- **Action planning**: Develop a plan of action to address the feedback, which may include design changes, content adjustments, or functional enhancements.

- **Implementation and testing**: Implement the necessary changes and conduct thorough testing to ensure that the modifications effectively address the feedback without introducing new issues.

- **Iteration**: Recognize that this is an iterative process. Continuous monitoring and further feedback collection are essential for ongoing improvement.

Communicating the changes made in response to user feedback is essential for maintaining trust and engagement with your user base. This can be achieved through update logs on the website, emails to users who provided feedback, or posts on social media channels. Transparent communication underscores a commitment to user satisfaction and can turn users into advocates for the brand.

Finally, measuring the impact of changes made in response to user feedback on the identified KPIs is critical for validating the effectiveness of the integration process. Improvement in user engagement metrics, conversion rates, or customer satisfaction scores can directly reflect the success of integrating user feedback into QA processes.

Integrating user feedback into QA processes for RWD projects is not just beneficial; it's also essential for creating websites that meet and exceed user expectations. By directly linking QA efforts with user insights, businesses can ensure that their RWD initiatives are genuinely user-centric, driving both user satisfaction and business success.

QA strategies to enhance user engagement and retention

QA strategies play a pivotal role in enhancing user engagement and retention, especially within the framework of RWD. The goal of QA in this context extends beyond ensuring that a website is free from bugs and errors; it encompasses creating an engaging, seamless, and rewarding user experience across all devices. By focusing on the end-user experience, QA processes can significantly contribute to achieving key business outcomes such as increased user loyalty, higher engagement rates, and improved conversion metrics.

Here are some strategies to achieve that:

- User engagement starts with performance. Websites that load quickly and perform smoothly across various devices and network conditions stand a better chance of retaining users. QA strategies should, therefore, include comprehensive performance testing, covering aspects such as load times, response times, and the efficiency of **content delivery networks** (**CDNs**). Optimizing images, leveraging browser caching, and minimizing the use of blocking JavaScript and CSS in above-the-fold content are critical tasks that directly impact performance and user satisfaction.

- Accessibility is a critical component of user engagement and retention. Websites that are accessible to all users, including those with disabilities, are more likely to foster positive user experiences and build loyalty. QA processes must incorporate accessibility testing, ensuring compliance with standards such as WCAG. This includes verifying color contrast, keyboard navigability, screen reader compatibility, and the proper use of **Accessible Rich Internet Applications** (**ARIA**) landmarks and roles.

- A core aspect of RWD is ensuring that websites are not only accessible but also usable across a wide range of devices. QA teams play a crucial role in validating the usability of websites by conducting cross-device testing to ensure consistent navigation, readability, and interactive elements. This involves testing on actual devices or using device emulators to replicate various user environments. Usability testing should focus on ease of navigation, clarity of content, and the intuitiveness of interactive elements, with a keen eye on the unique requirements of mobile users.

- A dynamic QA strategy involves not just testing but also learning from real user interactions. Implementing continuous feedback mechanisms — such as user surveys, feedback forms, and analytics tracking — enables you to collect valuable insights into user behavior and preferences. Analyzing this data helps identify areas for improvement, guiding QA efforts to focus on adjustments that enhance user engagement and retention.

- QA strategies should aim to streamline the user journey to reduce friction points that could lead to user drop-offs. This involves detailed testing of conversion paths, checkout processes, and form submissions to identify and eliminate obstacles that users may encounter. Simplifying these processes, providing clear **calls-to-action** (**CTAs**), and ensuring that the user journey is as intuitive as possible on all devices are key factors in enhancing user engagement.

- A/B testing is an invaluable tool for QA teams to empirically determine the impact of changes on user engagement and retention. By presenting two versions of a web page to users and analyzing the performance of each, QA teams can make data-driven decisions about design changes, content variations, and new features. A/B testing provides a scientific basis for enhancing the user experience and can significantly contribute to improved business outcomes.

In conclusion, QA strategies that focus on performance optimization, cross-device usability, accessibility, continuous feedback, streamlined user journeys, and empirical testing are essential for enhancing user engagement and retention in RWD projects. By aligning QA goals with these strategic areas, businesses can ensure that their web presence is not only technically sound but also deeply resonant with their user base, driving loyalty, satisfaction, and, ultimately, business success.

Now, let's break down how we can ensure we are continuously improving.

Continuous feedback and iteration

In our fast-paced digital landscape, establishing a framework for continuous improvement is essential for maintaining and enhancing the quality and performance of RWD projects. This framework enables organizations to adapt to changing user needs, technological advancements, and competitive pressures. It involves creating structured yet flexible processes that facilitate ongoing feedback, analysis, and iterative enhancements, ensuring that digital assets remain effective, engaging, and aligned with business goals.

Here's the list of steps to follow to ensure continuous improvement:

1. **Define objectives and KPIs**: The first step in establishing a framework for continuous improvement is to clearly define the objectives of your RWD project and identify the relevant KPIs. Objectives should be aligned with broader business goals, such as improving user engagement, increasing conversion rates, or enhancing brand perception. KPIs might include metrics such as page load times, bounce rate, mobile conversion rate, and user satisfaction scores. These metrics will serve as benchmarks to evaluate progress and identify areas for improvement.

2. **Implement mechanisms to gather feedback**: Continuous improvement relies on a steady stream of feedback from various sources. Implement mechanisms to gather both quantitative data, such as web analytics and user interaction logs, and qualitative feedback, such as user surveys, feedback forms, and usability testing results. Social media platforms and customer support interactions can also provide valuable insights into user experiences and expectations. The goal is to capture a comprehensive view of how users interact with your RWD project across different devices.

3. **Analyze feedback and identify improvement opportunities**: Regularly analyze the feedback and data you collect to identify patterns, trends, and specific issues affecting the user experience and performance. Use this analysis to pinpoint opportunities for improvement, prioritizing them based on their potential impact on user satisfaction and business outcomes. This step may involve cross-functional collaboration between designers, developers, QA professionals, and business analysts to ensure a holistic understanding of the findings.

4. **Plan and implement iterative enhancements**: Based on the analysis, develop a plan to implement iterative enhancements to your RWD project. This plan should outline the changes to be made, the resources required, and the timeline for execution. It's crucial to adopt an agile approach, allowing for flexibility and rapid response to insights gathered from ongoing feedback. Each iteration should be designed as a manageable cycle of development, testing, and deployment, enabling continuous refinement and optimization.

5. **Monitor the impact of changes**: After implementing enhancements, closely monitor their impact on the predefined KPIs and overall user experience. This monitoring should be both immediate, catching any potential issues introduced by the changes, and long-term, assessing their effectiveness over time. Use A/B testing or control groups to isolate the effects of specific changes, providing clear evidence of their value.

6. **Foster a culture of continuous improvement**: Perhaps the most critical element of a framework is fostering a culture of continuous improvement within an organization. Encourage open communication, collaboration, and experimentation across teams. Celebrate successes, learn from failures, and maintain a persistent focus on enhancing the user experience and achieving business goals. This cultural shift ensures that continuous improvement becomes an integral part of your organization's approach to RWD and digital strategy.

By establishing a framework for continuous improvement, organizations can ensure that their RWD projects remain dynamic, user-centric, and business-aligned. This iterative process not only enhances the quality and effectiveness of digital assets but also drives innovation and competitive advantage in the digital marketplace.

Leveraging analytics for data-driven iteration

In the realm of QA for RWD, leveraging analytics for data-driven iteration stands as a cornerstone strategy. This approach involves systematically using data gleaned from web analytics tools to inform decisions about design adjustments, feature enhancements, and overall user experience improvements. By closely monitoring user interactions, behaviors, and conversion metrics, teams can identify patterns and opportunities for optimization that are grounded in actual user data, rather than assumptions or general best practices.

The integration of analytics into the QA process begins with the selection of appropriate tools and metrics that align with a project's goals. Tools such as Google Analytics and Adobe Analytics offer a wealth of data on user behavior, including page views, session duration, bounce rates, and conversion paths. For RWD projects, it's crucial to segment this data by device type to understand how users interact with a site across different platforms.

To effectively leverage analytics, it's important to identify which metrics will serve as key indicators of performance and user satisfaction. These may include the following:

*   **Page load time**: Especially critical for mobile users, where delays can lead to increased bounce rates

*   **Bounce rate**: High bounce rates may indicate issues with content relevance, usability, or performance

*   **Conversion rate**: By device type, pinpointing where a design may fall short in guiding users to conversion

*   **User flow**: Understanding how users navigate through a site can reveal friction points in the user journey

With key metrics identified, the next step is to dive deep into the data to extract actionable insights. This involves looking for trends over time, comparing user behavior across devices, and identifying pages or features with suboptimal performance. Advanced techniques such as cohort analysis and user segmentation can provide deeper insights into specific user groups or behaviors.

Armed with data-driven insights, a team can prioritize areas for improvement and begin implementing changes iteratively. This might involve A/B testing different design elements, optimizing page load times, or refining the mobile user experience based on specific insights about how mobile users interact with a site. The goal is to make informed, incremental improvements that can be directly linked back to the analytics data.

After implementing changes, it's essential to measure their impact on the identified metrics. This not only validates the effectiveness of the adjustments but also provides a feedback loop for further iteration. If the changes lead to improvements in user engagement, conversion rates, or other key metrics, this success can inform future iterations. Conversely, if the data does not show the expected improvements, it can prompt a reevaluation and further refinement.

Cultivating a data-driven culture within a team is critical for the success of this approach. Encouraging collaboration between designers, developers, and analysts ensures that decisions are made with a comprehensive understanding of the data. Regularly reviewing analytics together can foster a shared commitment to continuous improvement and optimization.

Leveraging analytics for data-driven iteration in RWD projects empowers teams to make informed decisions that directly enhance user experience and business outcomes. By grounding QA and design decisions in real user data, organizations can ensure their responsive websites not only meet but also exceed user expectations, driving engagement, satisfaction, and conversion in an ever-evolving digital landscape.

The role of user testing in refining RWD

The role of user testing in refining RWD is integral to ensuring that websites not only adapt to various screen sizes but also deliver a seamless and engaging user experience across all devices. User testing provides direct insights into how real users interact with a website, uncovering usability issues, content clarity problems, and navigational challenges that might not be evident through analytics or internal review alone. By incorporating user testing into the continuous feedback and iteration cycle, teams can make data-driven decisions that significantly enhance the effectiveness and user satisfaction of their RWD projects.

Integrating user testing into the RWD process involves planning and executing tests at various stages of the project life cycle. Early-stage testing, such as wireframe or prototype testing, can reveal conceptual or structural issues before significant development efforts are undertaken. Later, live site testing with more refined implementations can uncover more nuanced usability issues or device-specific problems.

Several user testing methods can be applied, depending on the project phase and the specific insights needed:

- **Remote usability testing**: Allows testers from various locations to use a website in their natural environment, offering insights into real-world usability across devices

- **In-person testing**: Facilitates a deeper understanding of user behavior and provides immediate feedback, allowing for follow-up questions to probe deeper into the user experience

- **A/B testing**: Compares two versions of a web page to see which performs better on specific metrics, offering empirical evidence about the effectiveness of design choices

- **Heatmaps and click tracking**: Provides visual representations of how users interact with a site, indicating which areas attract the most attention and interaction

The data gathered from user testing needs to be carefully analyzed to identify patterns and key takeaways. It's important to prioritize issues based on their impact on the user experience and the business goals of the website. Addressing critical usability issues, especially those that directly affect conversion rates or user engagement, should take precedence.

After identifying the necessary changes, the design and development teams should collaborate to implement solutions. This might involve redesigning elements for better usability, adjusting layouts to improve content visibility on mobile devices, or refining interactive features to ensure that they are intuitive across all platforms.

User testing for RWD is inherently iterative. It's not a one-time task but an ongoing process that evolves with a website. As new features are added, or as user expectations change, regular user testing sessions can help ensure that the website continues to meet and exceed user needs. Each round of testing provides fresh insights, feeding into a cycle of continuous improvement.

Incorporating regular user testing into the RWD process fosters a culture of user-centric design within an organization. It emphasizes the importance of understanding and meeting user needs, encouraging teams to view the website from the perspective of its users. This shift in mindset is crucial for creating digital experiences that are not only technically responsive but also deeply resonant with the target audience.

In conclusion, the role of user testing in refining RWD is critical. It bridges the gap between technical responsiveness and actual user satisfaction, ensuring that websites are not only adaptable to different devices but are also intuitive, engaging, and effective in meeting user needs. Now, we need to explore how we accurately measure the ROI for our RWD efforts.

Measuring the tangible ROI of responsive initiatives

Integrating user testing into the RWD process involves planning and executing tests at various stages of the project life cycle. Early-stage testing, such as wireframe or prototype testing, can reveal conceptual or structural issues before significant development efforts are undertaken. Later, live site testing with more refined implementations can uncover more nuanced usability issues or device-specific problems.

Several user testing methods can be applied, depending on the project phase and the specific insights needed:

- **Remote usability testing**: Allows testers from various locations to use the website in their natural environment, offering insights into real-world usability across devices

- **In-person testing**: Facilitates a deeper understanding of user behavior and provides immediate feedback, allowing for follow-up questions to probe deeper into user experiences

- **A/B testing**: Compares two versions of a web page to see which performs better on specific metrics, offering empirical evidence about the effectiveness of design choices

- **Heatmaps and click tracking**: Provides visual representations of how users interact with the site, indicating which areas attract the most attention and interaction

The data gathered from user testing needs to be carefully analyzed to identify patterns and key takeaways. It's important to prioritize issues based on their impact on the user experience and the business goals of the website. Addressing critical usability issues, especially those that directly affect conversion rates or user engagement, should take precedence.

After identifying the necessary changes, the design and development teams should collaborate to implement solutions. This might involve redesigning elements for better usability, adjusting layouts to improve content visibility on mobile devices, or refining interactive features to ensure they are intuitive across all platforms.

User testing for RWD is inherently iterative. It's not a one-time task but an ongoing process that evolves with a website. As new features are added, or as user expectations change, regular user testing sessions can help ensure that the website continues to meet and exceed user needs. Each round of testing provides fresh insights, feeding into a cycle of continuous improvement.

Incorporating regular user testing into the RWD process fosters a culture of user-centric design within an organization. It emphasizes the importance of understanding and meeting user needs, encouraging teams to view a website from the perspective of its users. This shift in mindset is crucial for creating digital experiences that are not only technically responsive but also deeply resonant with the target audience.

In conclusion, the role of user testing in refining RWD is critical. It bridges the gap between technical responsiveness and actual user satisfaction, ensuring that websites are not only adaptable to different devices but are also intuitive, engaging, and effective in meeting user needs. By embedding user testing into the continuous feedback and iteration cycle, teams can achieve a truly responsive and user-centered design.

Quantitative metrics to evaluate success

Quantitative metrics play a crucial role in evaluating the success of RWD initiatives, offering tangible data that can directly inform the ROI calculation. These metrics provide clear, measurable insights into how RWD affects user behavior, website performance, and ultimately, business outcomes. By closely monitoring a set of KPIs, organizations can assess the effectiveness of their RWD projects and identify areas for further optimization.

Let's look at the metrics:

- **Traffic metrics**:

  - **Mobile traffic volume**: An increase in mobile traffic post-RWD implementation indicates success in attracting mobile users

  - **Traffic by device type**: Segregating traffic by device type (mobile, tablet, or desktop) helps to understand user preferences and optimize the design accordingly

- **Engagement metrics**:

  - **Bounce rate by device**: A decrease in bounce rates, especially on mobile devices, suggests that users find the responsive site more engaging.

  - **Average session duration and pages per session**: Increases in these metrics post-RWD rollout indicate that users find the content more accessible and compelling across devices

- **Conversion metrics**:

  - **Conversion rate by device type**: Post-RWD improvements in conversion rates, particularly on mobile and tablet devices, directly correlate to the ROI, showing that the responsive design effectively drives desired user actions

  - **Mobile conversion growth**: A significant metric for businesses, especially with the increasing prevalence of mobile commerce

- **Performance metrics**:

  - **Page load time**: Faster load times across devices post-RWD implementation can significantly enhance user satisfaction and contribute to higher engagement and conversion rates

  - **Website downtime**: Reduced downtime across devices signifies a stable and reliable user experience, contributing to overall business success

- **SEO metrics**:

  - **Organic search rankings**: Improvements in search rankings, particularly for mobile searches, can indicate the success of RWD in meeting search engine guidelines for mobile-friendliness.

  - **Organic traffic**: An increase in organic traffic post-RWD reflects higher visibility and improved SEO performance

- **Financial metrics**:

  - **Revenue Per Visitor (RPV) by device**: An increase in RPV following RWD implementation demonstrates direct financial benefits, showing that users are spending more

  - **Cost Per Acquisition (CPA) by device**: Monitoring changes in CPA can help assess the cost-effectiveness of RWD in attracting and converting users.

- **User satisfaction metrics**:

  - **Net Promoter Score (NPS)**: While more qualitative, an improvement in NPS can provide a quantifiable measure of user satisfaction and loyalty post-RWD

  - **Customer feedback and reviews**: Increases in positive feedback and higher ratings on mobile usability can serve as indirect quantitative metrics to evaluate RWD success

Next, let's delve into monitoring and analysis.

Monitoring and analysis

Continuous monitoring and detailed analysis of these metrics are essential for accurately measuring the ROI of RWD initiatives. Tools such as Google Analytics offer comprehensive capabilities to track these KPIs, allowing businesses to drill down into the data and gain actionable insights. It's important to compare pre- and post-RWD launch metrics to assess the direct impact of responsive design on website performance and user engagement.

In conclusion, leveraging these quantitative metrics offers a systematic approach to evaluating the success of RWD projects. By aligning these metrics with specific business objectives, organizations can not only justify their investment in RWD but also continue to refine their digital strategies for optimal performance and ROI.

Qualitative benefits and their business impact

While quantitative metrics are crucial for measuring the tangible ROI of RWD initiatives, qualitative benefits also play a significant role in determining overall business impact. These benefits, although not always directly measurable in numbers, contribute to long-term success by enhancing brand reputation, improving customer loyalty, and fostering a positive user experience. Recognizing and understanding the qualitative impact of RWD can help businesses appreciate the full spectrum of advantages that responsive design brings to the digital landscape.

Let's discuss some of the qualitative benefits.

RWD fundamentally aims to provide a seamless and consistent experience across all devices. This focus on UX can significantly reduce user frustration associated with poor navigation, slow loading times, and non-intuitive site layouts. While specific UX improvements may not always translate directly into immediate measurable returns, they lay the foundation for increased user satisfaction, higher engagement levels, and the potential for future conversions.

A responsive website reflects a brand's commitment to catering to its customers' needs, showcasing the brand as modern, relevant, and customer-focused. This positive brand perception can lead to increased brand loyalty, as users are more likely to return to a site that offers a hassle-free browsing experience regardless of the device used. Over time, this loyalty translates into repeat business, referrals, and a strong, loyal customer base, driving long-term revenue growth.

In markets where competitors have yet to adopt RWD, offering a responsive site can provide a significant competitive edge. By meeting users' expectations for a fluid online experience across devices, businesses can capture market share from competitors whose websites are less accessible or user-friendly. This advantage is particularly pronounced in industries where mobile usage is high, and customer expectations for digital experiences are elevated.

By optimizing for mobile and other devices, RWD expands a website's reach to a wider audience. This inclusivity not only captures users who exclusively or primarily use mobile devices but also accommodates users with disabilities or those using alternative web browsing technologies. Expanding the potential audience in this way can lead to market expansion and the tapping of previously underserved or unrecognized segments.

A qualitative benefit of RWD is its role in mitigating risks associated with non-compliance with web standards and accessibility guidelines. By ensuring a website is responsive and accessible, businesses can avoid potential legal complications and penalties associated with failing to meet these standards. Furthermore, RWD helps safeguard against the risk of obsolescence, ensuring the website remains functional and relevant as new devices and browsing methods emerge.

Implementing RWD can also have internal benefits, such as boosting employee morale and aiding in recruitment. Teams that work on forward-thinking projects such as RWD often experience higher job satisfaction, due to the innovative nature of the work and the clear impact of their efforts on UX. Additionally, businesses known for their commitment to modern web practices may find it easier to attract top talent in the design, development, and UX fields.

While these qualitative benefits may not directly translate into immediate financial gains, they contribute significantly to the overall success and sustainability of a business. Over time, qualitative benefits such as enhanced UX, brand loyalty, and competitive advantage can lead to quantifiable outcomes, such as increased sales, lower customer acquisition costs, and higher lifetime value of customers.

So, the qualitative benefits of RWD initiatives have a profound business impact that complements the quantitative ROI. Together, they provide a comprehensive view of the value that responsive design brings to a business, underscoring the importance of considering both qualitative and quantitative factors in evaluating the success of RWD projects.

Summary

Throughout this chapter, we've embarked on an in-depth exploration of the vital role that QA plays in the realm of RWD, highlighting the necessity of implementing specialized testing strategies tailored to the multifaceted nature of responsive projects. We've dissected the methodology behind rigorous QA processes, emphasizing their indispensable contribution to delivering high-quality, user-centric web experiences that perform seamlessly across a spectrum of devices and screen sizes.

Key lessons in this chapter have equipped you with an understanding of how to conduct effective cross-device testing, ensuring that every user enjoys an optimal browsing experience, regardless of how they access your site. We also delved into the strategic measurement of ROI for RWD projects, demonstrating how to blend quantitative metrics with qualitative insights to gauge the success and impact of your responsive initiatives accurately. This knowledge is crucial for justifying investments in RWD and guiding future enhancements and optimizations.

Moreover, we addressed the critical balance between maintaining QA rigor and adhering to project timelines — a challenge that demands thoughtful planning, prioritization, and agility. This section underscored the importance of adopting efficient workflows and leveraging advanced tools to streamline the QA process, without compromising the depth or quality of testing.

Understanding these principles is immensely beneficial, not only for ensuring the technical and functional excellence of RWD projects but also for aligning them closely with business goals and user expectations. The skills learned in this chapter are foundational to building digital solutions that are not just responsive but also robust, engaging, and strategically aligned with long-term business objectives.

As we move on to the next chapter, we'll build on what we've established here by exploring how the choices of tools, platforms, and vendor partnerships can significantly influence the success of responsive projects. This next step is a natural progression from ensuring quality and measuring ROI, as the platforms and technologies you select are pivotal in executing your RWD strategy efficiently and effectively.

12
Vendor and Platform Decisions in RWD

In the rapidly evolving world of **responsive web design** (**RWD**), the selection of platforms and vendors plays a critical role in the success of any project. This chapter dives into the essential considerations and strategies for making informed decisions in this crucial area. It will equip you with the skills to evaluate and select the best platforms and tools tailored to the unique demands of RWD projects, ensuring that your digital assets are not only responsive but also scalable and future-proof.

As the digital landscape continues to shift, understanding the business implications of these platform choices becomes paramount. From open source frameworks to proprietary solutions, each option offers distinct advantages and challenges that can significantly impact project timelines, budget, and overall user experience. This chapter aims to demystify these options, providing clear guidance on assessing each platform's strengths and limitations.

Furthermore, effective management of vendor relationships is essential for the smooth execution of RWD projects. Navigating these partnerships with clarity and strategic insight can enhance collaboration, ensuring that all parties are aligned with the project's goals and timelines. You will learn the best practices for fostering productive vendor relationships that support the creative and technical demands of responsive design.

Lastly, this chapter delves into the importance of assessing platform flexibility and future scalability. In an era where technological advancements are constant, choosing platforms that can adapt to emerging trends and user expectations is vital. Through real-world examples and expert insights, you will discover how to make forward-looking decisions that accommodate growth and innovation, ensuring your RWD efforts stand the test of time.

This chapter is designed to provide a comprehensive understanding of the factors that influence platform and vendor selection, empowering you to make choices that align with your strategic goals and enhance your ability to create engaging and responsive digital experiences.

The following topics will be covered in this chapter:

- Evaluating vendor capabilities for RWD
- Open source versus proprietary platforms — a strategic choice
- Platform adaptability and future growth
- Navigating the vendor partnership landscape

Evaluating vendor capabilities for RWD

In the evolving landscape of RWD, selecting the right vendor is pivotal to the success of your digital strategy. Understanding the fundamentals of vendor evaluation is the first step in forging partnerships that align with your project's objectives and long-term digital goals. This chapter delves into the core considerations for assessing potential RWD vendors, providing a framework to ensure that your choice not only meets your current requirements but also supports future scalability and innovation.

Before embarking on the vendor evaluation process, clearly define what you need from your RWD project. Consider aspects such as the scope of the project, target audience, **key performance indicators** (**KPIs**), and specific features or functionalities that are crucial for your website. Having a clear understanding of your requirements will allow you to assess vendors more effectively, focusing on those who demonstrate strength in areas most critical to your project's success.

A vendor's technical proficiency in RWD is fundamental. You must evaluate their understanding and application of fluid grids, flexible images, and media queries as these are the technical cornerstones of responsive design. Inquire about the frameworks and coding standards they employ and consider their experience with accessibility guidelines and cross-browser compatibility. A portfolio of successful RWD projects can serve as a testament to their expertise and give you insight into their design and development capabilities.

Understanding a vendor's approach to RWD projects is crucial. Discuss their strategy for mobile-first design versus desktop-first, and how they plan to integrate your website's content with the responsive layout. Their methodology for project management, from initial concepts to final testing, can reveal much about their organizational skills and ability to meet deadlines. Effective communication throughout the project life cycle is essential for a successful partnership.

Post-launch support and maintenance are critical aspects of a vendor's offerings. Responsive sites require ongoing optimization and updates to ensure they continue to perform well across all devices and browsers. Assess the vendor's commitment to post-launch support, including their process for handling bugs, their availability for future updates, and their approach to training your team to manage the website's day-to-day operations.

The vendor's corporate culture and approach to client relationships can significantly impact the success of your project. A vendor that values collaboration and transparency and has a client-centric approach is more likely to understand your vision and work diligently to achieve it. Establishing a strong working relationship based on mutual respect and shared goals is essential for navigating the challenges and changes that inevitably arise in RWD projects.

Armed with a comprehensive evaluation framework, you can make an informed decision that goes beyond cost considerations to select a vendor whose capabilities, approach, and values align with your RWD project goals. Remember, the right vendor partnership is not just about completing a project; it's about setting the foundation for your digital presence to adapt and thrive in an ever-changing technological landscape.

Next, we need to explore how to best assess RWD expertise.

Key criteria for assessing RWD expertise

Evaluating a vendor's expertise in RWD requires a deep dive into their technical skills, project approach, and the outcomes they've achieved for other clients. The complexity of RWD projects, with their need for adaptability across various devices and screen sizes, makes it crucial to assess a vendor's capabilities meticulously. Here are the key criteria to consider when assessing a vendor's expertise in RWD:

- A fundamental criterion is the vendor's command over the technical aspects of RWD. This includes their ability to implement fluid grids, flexible images, and media queries effectively. Ask for examples of how they've utilized these techniques in past projects. Proficiency in frontend web technologies such as HTML5, CSS3, and JavaScript, along with frameworks such as Bootstrap or Foundation, helps with creating responsive designs that perform well across all devices.

- RWD is not just about making a website viewable on any device; it's about ensuring a seamless and engaging user experience regardless of the screen size. Evaluate the vendor's approach to design from a UX perspective. Do they prioritize mobile users? How do they plan navigation and layout changes across devices? Assess their understanding of touch versus mouse interactions and their ability to design accessible websites compliant with **Web Content Accessibility Guidelines (WCAG)**.

- Given that mobile users often rely on less stable network connections, a vendor's ability to optimize website performance is crucial. Inquire about their strategies for minimizing load times, such as image optimization, minifying CSS and JavaScript, and leveraging browser caching. Their understanding of performance metrics and tools for ongoing monitoring will also indicate their commitment to delivering high-quality RWD solutions.

- Understanding the vendor's process for executing RWD projects provides insight into their efficiency and reliability. Look for a vendor with a structured approach that includes phases such as discovery, design, development, testing, and deployment. Communication is key to a successful partnership; hence, assess their responsiveness, the clarity of their project updates, and their willingness to incorporate feedback.

- Responsive design requires continuous adjustments and optimizations based on user feedback and evolving web standards. A competent RWD vendor should offer robust post-launch support, including monitoring website performance, addressing any usability issues, and making necessary design or content adjustments. This ongoing support is vital for keeping the website relevant and engaging over time.

By evaluating vendors against these criteria, businesses can select a partner with the proven RWD expertise necessary to create a website that not only looks great on any device but also offers a superior user experience, drives engagement, and, ultimately, enhances the brand's online presence.

Now, we need to ensure that we understand a crucial difference in RWD between open source and proprietary platforms.

Open source versus proprietary platforms — a strategic choice

In the realm of RWD, choosing between open source and proprietary platforms is a critical strategic decision. This choice can significantly impact your project's flexibility, cost, and long-term viability. Understanding the fundamental differences between these two types of platforms is essential for making an informed decision that aligns with your business goals and technical requirements.

Open source platforms

Open source platforms are characterized by their freely available source code that anyone can inspect, modify, and enhance. These platforms are developed and maintained by a community of contributors and often come with no licensing fees, making them an attractive option for businesses looking to minimize upfront costs. However, we must consider their advantages and disadvantages.

Here is a list of the advantages of open source platforms:

- **Cost-effectiveness**: Most open source platforms are free to use, offering a lower total cost of ownership compared to proprietary solutions

- **Flexibility and customization**: With access to the source code, developers can customize the platform to meet specific project requirements without the constraints imposed by proprietary systems

- **Community support**: Open source projects benefit from the collective knowledge and contributions of a global community, offering extensive resources, plugins, and themes that enhance the platform's capabilities

The following is a list of the disadvantages of open source platforms:

- **Resource intensity**: The need for customization and the potential lack of official support can require more time and expertise, potentially increasing the total project cost
- **Security concerns**: While open source communities work diligently to identify and patch vulnerabilities, the public nature of the code can expose it to security risks

Now, let's consider the same for proprietary platforms.

Proprietary platforms

Proprietary platforms are owned by a single company that controls the software's source code. Users of proprietary software typically pay for a license to use the platform, which includes customer support and updates provided by the owning company.

Here is a list of the advantages of proprietary platforms:

- **Streamlined support**: Proprietary platforms often come with dedicated support and regular updates, ensuring users have assistance when needed and the platform remains secure and up to date
- **Ease of use**: These platforms are usually designed with user-friendliness in mind, offering intuitive interfaces and functionalities that can reduce the time and expertise required to launch and manage a website

The following is a list of the disadvantages of proprietary platforms:

- **Cost**: Licensing fees, subscriptions, and costs for additional features or updates can make proprietary platforms more expensive over time
- **Limited customization**: Users are generally limited to the features and customizations allowed by the platform, which can be restrictive for businesses with unique or evolving needs

Now that we understand the advantages and disadvantages of both platforms, let's look at how we can make the right choice.

Making a strategic choice

The decision between an open source and a proprietary platform hinges on several factors, including budget, project complexity, available expertise, and long-term digital strategy. While open source platforms offer unparalleled customization and cost benefits, they require a hands-on approach and a clear understanding of the potential security and maintenance demands. Proprietary platforms, on the other hand, provide streamlined support and ease of use at a cost, which might be justifiable for businesses seeking reliability and simplicity in their RWD endeavors.

Ultimately, defining your project's specific needs, constraints, and goals is essential for choosing the platform that will best support your RWD project now and in the future.

Now, let's ensure that we are bulletproofing ourselves for the future.

Platform adaptability and future growth

In the dynamic realm of RWD, understanding the spectrum of testing methods is paramount for ensuring that websites deliver optimal user experiences across a multitude of devices. RWD testing is not a one-size-fits-all process; it encompasses a variety of tests, each designed to address specific aspects of the website's functionality, appearance, and performance on different screen sizes and resolutions. This comprehensive approach to testing is crucial for identifying and resolving issues that could detract from the user experience or hinder the site's performance.

Choosing a platform for your RWD project involves not just assessing its current capabilities but also considering its scalability and adaptability for future growth. As your business evolves, your web platform must be capable of accommodating increased traffic, expanding content, and integrating new functionalities without compromising performance or user experience. Next, we will learn how to assess a platform's scalability for your RWD projects.

Understanding scalability in the context of RWD

Scalability refers to a platform's ability to handle growing amounts of work or its potential to be enlarged to accommodate that growth. For RWD, this means the platform should support increasing numbers of users, devices, and interactions while maintaining fast loading times and a consistent experience across all screen sizes.

Here are the key factors to consider for enhanced scalability:

- **Performance under load**: Evaluate how the platform performs under increased traffic. Can it maintain speed and reliability when user numbers spike? Platforms that can dynamically allocate resources based on demand are typically more scalable.

- **Content management flexibility**: As your site grows, you'll likely add more content. A scalable RWD platform should offer efficient content management capabilities that allow for easy updates and expansions without the need for extensive backend overhauls.

- **Integration capabilities**: The ability to integrate with other tools and services (for example, CRM systems, e-commerce platforms, and social media) is crucial for scalability. A platform that supports APIs and has a robust ecosystem of plugins and extensions can adapt more readily to changing business needs.

- **Customization and extension**: Assess whether the platform allows for custom development. Being able to add custom features or extend existing functionalities without significant rework is vital for adapting to future trends and user expectations.

Now, let's investigate how to assess future growth potential.

Assessing future growth potential

When thinking about the potential of our platform to grow, there are a couple of key things to consider:

- **Vendor roadmap**: Look into the platform vendor's development roadmap. Does it align with web standards and emerging technologies? Platforms committed to continuous improvement are more likely to support your long-term growth.

- **Community and support**: For open source platforms, the strength and activity of the developer community can be a good indicator of its scalability and adaptability. A vibrant community means regular updates, security patches, and a wide range of plugins or themes.

- **Case studies and references**: Review case studies or testimonials from businesses that have scaled their operations using the platform. Learning from others' experiences can provide insights into potential challenges and benefits.

- **Technical support**: Consider the level of technical support offered, especially for proprietary platforms. As your site grows, having access to responsive, knowledgeable support can be invaluable for resolving issues quickly.

Assessing a platform's scalability involves comprehensively evaluating its ability to grow with your business, adapt to technological advancements, and maintain high performance and usability standards. By prioritizing scalability in your platform decision, you can ensure that your RWD project remains robust and responsive to the changing digital landscape, safeguarding your investment and supporting your business's future success.

Now, let's explore how we can ensure that we remain agile by consistently adapting to the changing landscape.

Adaptability to emerging web technologies

In the rapidly evolving digital landscape, the adaptability of your chosen platform to emerging web technologies is a crucial factor for the sustainability and growth of your RWD initiatives. This adaptability ensures that your website remains at the forefront of technological advancements, providing optimal user experiences and leveraging the latest innovations for enhanced performance and functionality.

Understanding adaptability in web platforms

Adaptability in the context of web platforms refers to the ease with which a platform can incorporate new web standards, technologies, and user expectations. It encompasses the platform's ability to evolve, supporting new features and integrations without requiring a complete overhaul of the existing system.

Why adaptability matters

Let's look at why adaptability matters:

- **Keeping pace with web standards**: The web is governed by constantly evolving standards aimed at improving accessibility, performance, and security. A platform that adapts to these changes ensures your website remains compliant and functional across all browsers and devices.

- **Embracing new technologies**: As new technologies emerge, from **progressive web apps (PWAs)** to **artificial intelligence (AI)** and beyond, adaptability allows your RWD project to incorporate these innovations, enhancing capabilities and staying competitive.

- **Future-proofing your investment**: Choosing a platform that is adaptable to future web technologies protects your investment by extending the lifespan of your website and minimizing the need for frequent, costly rebuilds.

Now that we know why it matters, let's explore how to evaluate a platform's adaptability.

Evaluating a platform's adaptability

The following are some ways you can best evaluate a platform's adaptability:

- **Modular architecture**: Platforms with a modular architecture allow for easier updates and additions. This flexibility is key to integrating new technologies and features as they become available.

- **Active development and community support**: For open source platforms, an active development community is a good indicator of adaptability. A committed community and regular updates suggest that the platform will continue to evolve in line with technological advancements.

- **Documentation and learning resources**: Comprehensive documentation and accessible learning resources are indicators of a platform's commitment to adaptability. They provide developers with the guidance needed to implement new technologies and best practices.

- **API support and ecosystem**: A platform with robust API support and a rich ecosystem of plugins, extensions, or add-ons is more adaptable. These features allow for seamless integration with emerging technologies and third-party services.

Now, let's explore how we can use our knowledge of platform adaptability to make informed, strategic decisions.

Making strategic decisions

When assessing a platform's adaptability, consider not only its current capabilities but also its strategic direction and how it aligns with future web trends. Engage with vendors or the developer community to understand the platform's roadmap and how it plans to embrace emerging technologies.

Incorporating adaptability into your platform selection criteria ensures that your RWD project remains flexible, relevant, and capable of evolving. This forward-thinking approach is essential for businesses aiming to leverage digital channels effectively, providing a foundation that supports growth, innovation, and continued engagement with your audience in the ever-changing digital environment.

RWD, just like everything else in technology, is going to rapidly change and evolve. Let's make sure that your strategy is ready to adapt and evolve with it.

Future-proofing your RWD strategy

Future-proofing your RWD strategy is essential for maintaining an effective, engaging online presence that can adapt to technological advancements and changing user behaviors over time. This process involves selecting a platform and adopting practices that ensure your website remains relevant, accessible, and performance-driven, regardless of how digital landscapes evolve. So, how can you future-proof your RWD strategy? Let's take a look:

- **Emphasize scalable and flexible design:**

  - **Scalable architecture**: Choose a platform with a scalable architecture that can handle increasing traffic and content without degrading performance. This ensures that as your business grows, your website can grow with it, accommodating more users, pages, and features seamlessly.

  - **Flexible content management**: Opt for a **content management system (CMS)** that allows you to easily update and reconfigure content layouts without needing extensive redevelopment. A flexible CMS is crucial for adapting your content strategy to meet future demands and user expectations.

- **Incorporate best practices for longevity:**

  - **Standards-compliant coding**: Use clean, standards-compliant HTML, CSS, and JavaScript to build your site. Adherence to web standards ensures compatibility with future web browsers and devices, making it easier to update your site as technologies change.

  - **Mobile-first approach**: A mobile-first design philosophy not only addresses the immediate needs of the growing number of mobile users but also positions your website to be more adaptable to new screen sizes and interaction models that may emerge in the future.

  - **Modular design and development**: Implementing a modular approach to both design and development allows individual components of your website to be updated or replaced without impacting the overall system. This modularity facilitates easier integration of new technologies and features over time.

- **Stay ahead with continuous learning and innovation**:

  - **Ongoing education and training**: Encourage continuous learning within your team. Staying informed about emerging web design trends, technologies, and user experience strategies is key to identifying opportunities for innovation within your RWD projects.

  - **Experimentation and testing**: Foster a culture of experimentation, where new ideas and technologies can be tested and iterated upon. Use A/B testing and user feedback to guide the evolution of your website, ensuring it remains aligned with user needs and preferences.

- **Evaluate and adapt to technological advancements**:

  - **Regular platform assessments**: Conduct regular reviews of your web platform and infrastructure to identify areas that may require updates or optimizations. This includes assessing the performance, security, and mobile responsiveness of your site.

  - **Integration of emerging technologies**: Keep an eye on technological advancements that could enhance your RWD strategy, such as AI, voice search optimization, or **augmented reality** (**AR**). Assess how these technologies could be integrated into your website to improve user engagement and stay ahead of competitors.

Future-proofing your RWD strategy requires a proactive approach to platform selection, design, and development practices, coupled with an ongoing commitment to adaptation and innovation. By building a scalable, flexible foundation and fostering a culture of continuous improvement, you can ensure that your website remains effective and engaging for years to come, regardless of how the digital landscape evolves.

Now, it's time to break down the importance of navigating the partnerships with vendors.

Navigating the vendor partnership landscape

Navigating the vendor partnership landscape for RWD demands a strategic approach to selecting the right vendor that aligns with your project's goals, budget, and timelines. The right partnership can significantly influence the success of your RWD initiative, impacting everything from design quality to functionality and overall user experience. Here are the essential criteria to consider when selecting the right RWD vendor:

- **Experience and expertise**: Look for vendors with a proven track record in delivering successful RWD projects. Experience across various industries and the complexities of different projects can be a testament to their adaptability and expertise. Evaluate their portfolio for examples of past work that aligns with your project's scope and complexity.

- **Technical proficiency**: Assess the vendor's technical skills in RWD principles, including fluid grids, flexible images, and media queries. Their proficiency in frontend development languages (HTML, CSS, JavaScript) and frameworks (Bootstrap, Foundation) is crucial. Additionally, their approach to performance optimization and accessibility standards indicates their capability to deliver high-quality, responsive sites.

- **Strategic approach to design**: A vendor should demonstrate a strategic approach to design, prioritizing user experience across devices. This includes a mobile-first or content-first strategy that ensures seamless navigation and interaction, regardless of the user's device. Their process should integrate UX research, persona development, and user testing to inform design decisions.

- **Project management and communication**: Effective project management and clear communication are vital for the timely and successful completion of an RWD project. Inquire about their project management tools, methodologies (for example, Agile, Waterfall, and so on), and communication protocols. Understanding how they handle project milestones, feedback loops, and revisions will give you insight into their operational efficiency.

- **Scalability and future growth**: Consider the vendor's ability to scale the project for future growth. This includes their approach to modular design, CMSs, and third-party integrations. The vendor should offer solutions that are not just viable for the present but can adapt to future technological advancements and business needs.

- **Support and maintenance**: Post-launch support is critical to the ongoing success of an RWD site. Discuss the vendor's policies on maintenance, updates, and troubleshooting. Understanding the level of support offered (for example, dedicated account manager, support tickets, and emergency contacts) will help you manage expectations and ensure long-term site performance.

- **Cost and value proposition**: While cost should not be the sole deciding factor, it's important to ensure the vendor's pricing aligns with your budget and provides value for the investment. Request detailed proposals that outline the scope of work, deliverables, timelines, and costs. Comparing proposals from multiple vendors can provide a broader perspective on the market rates and value offerings.

- **Cultural fit**: Lastly, the vendor's corporate culture and values should resonate with your own. A partnership built on mutual respect, shared goals, and a collaborative spirit fosters a productive working relationship. Meetings and discussions can offer glimpses into their team dynamics, responsiveness, and commitment to your project's success.

Selecting the right RWD vendor is a comprehensive process that involves evaluating technical skills, design approach, project management capabilities, and cultural fit. By carefully considering these criteria, you can establish a partnership that not only meets your current RWD needs but also supports your digital strategy in the long run.

Partnerships are crucial for the success of any business. Let's see how that can aid in RWD success.

Building effective partnerships for RWD success

Building effective partnerships for RWD success is about more than just selecting a vendor with the right technical skills and experience. It involves establishing a collaborative relationship that fosters communication, mutual understanding, and shared objectives. Here are some key strategies you should follow to ensure your partnership with an RWD vendor leads to the successful realization of your project goals:

- **Establish clear communication channels**: Open and ongoing communication is the cornerstone of any successful partnership. Define clear channels and protocols for communication at the outset of the project. Whether it's regular status meetings, email updates, or collaborative project management tools, ensure that there are agreed-upon methods for keeping all parties informed and engaged throughout the project life cycle.

- **Define roles and responsibilities**: Clarifying roles and responsibilities from the start helps prevent overlaps and gaps in project execution. Clearly outline what is expected from both the vendor and your internal team. Understanding who is responsible for each aspect of the project, from content creation to technical development and quality assurance, ensures a cohesive approach to the project.

- **Set shared goals and objectives**: Aligning on shared goals and objectives ensures that both parties are working toward the same vision of project success. Establishing these objectives early on, and revisiting them regularly, keeps the project focused and provides benchmarks for measuring progress and outcomes.

- **Foster collaboration and trust**: A partnership is most effective when there's a foundation of trust and collaboration. Encourage open dialog about challenges and opportunities and be open to suggestions and innovations from your vendor. Trusting your vendor's expertise and allowing them creative freedom within agreed parameters can lead to enhanced outcomes.

- **Implement Agile methodologies**: Adopting an Agile approach to project management can enhance flexibility and responsiveness to change, which is crucial in RWD projects. Agile methodologies encourage iterative development, regular feedback, and adaptive planning, making it easier to adjust to project demands and market trends as they evolve.

- **Focus on knowledge sharing**: Effective partnerships are also learning opportunities. Encourage knowledge sharing between your vendor and internal teams. This can involve technical training sessions, workshops on best practices in RWD, or regular insights into emerging trends. Knowledge sharing strengthens the capabilities of your team and enriches the partnership.

- **Plan for long-term engagement**: Consider the vendor partnership not just as a one-time project engagement but as a long-term relationship that can grow and evolve. Discuss plans for future phases, ongoing support, and maintenance from the outset. A vendor that is invested in your long-term success will be more proactive about delivering high-quality work and staying engaged post-launch.

- **Conduct regular reviews and feedback sessions**: Regular reviews and feedback sessions help assess the project's progress against goals and identify areas for improvement. Constructive feedback is valuable for both parties to refine their processes and deliverables. Celebrate successes and learn from challenges together to strengthen the partnership.

Building effective partnerships for RWD success requires deliberate efforts to establish clear communication, shared goals, and mutual trust. By focusing on collaboration, flexibility, and long-term engagement, you can create a productive relationship with your vendor that not only achieves your current project objectives but also positions you for future digital initiatives.

Managing and evaluating vendor performance

Managing and evaluating vendor performance is a crucial aspect of ensuring the success of your RWD projects. A systematic approach to performance management helps in maintaining project quality, meeting timelines, and ensuring that the partnership delivers on its promises.

Here's how to effectively manage and evaluate your RWD vendor's performance throughout the project life cycle:

- Begin by establishing clear, measurable performance metrics based on the project's objectives. These can include project deliverables, quality standards, adherence to timelines, responsiveness to feedback, and the effectiveness of communication. Make sure these metrics are agreed upon by both parties to ensure alignment.

- Schedule regular check-ins or status meetings with your vendor to review project progress, discuss challenges, and adjust plans as necessary. These sessions provide you with an opportunity to assess performance against the established metrics and address any issues before they escalate.

- Leverage project management tools that allow both parties to track progress, manage tasks, and share documents. Tools such as Trello, Asana, and Jira can offer transparency, facilitate communication, and help in keeping the project on track.

- Feedback should be timely, specific, and constructive. Highlight areas where the vendor is performing well to encourage good practices, and tactfully address areas that need improvement. Open and honest feedback is key to fostering a productive partnership.

- At major project milestones, conduct formal evaluations of the vendor's performance. Review the work that's been completed against the agreed-upon metrics and document the outcomes of the evaluation. These evaluations can inform future work and help in refining project strategies.

- Approach vendor performance management as a collaborative effort rather than a vendor-client dynamic. Working together to solve problems, adapt to project changes, and achieve shared goals can lead to more successful outcomes and a stronger partnership.

- Use the insights gained from managing and evaluating vendor performance to plan for continuous improvement. Discuss how processes, communication, and collaboration can be enhanced for future phases of the project or new projects.

- In cases of underperformance, it's important to address the issue directly with the vendor. Provide clear examples of where expectations were not met and work together to develop a plan for corrective action. It's crucial to maintain professionalism and focus on solutions rather than placing blame.

- Recognizing and celebrating successes strengthens the partnership and motivates all parties involved. Acknowledge the vendor's contributions to the project's successes and share these achievements with your teams.

Managing and evaluating vendor performance is an ongoing process that requires attention, communication, and collaboration. By establishing clear expectations, maintaining open lines of communication, and focusing on continuous improvement, you can ensure that your vendor partnership contributes positively to the success of your RWD initiatives.

Summary

As we conclude our exploration of this chapter, it's clear that the strategic choices we make in terms of platforms and vendors are foundational to the success of RWD projects. Throughout this chapter, we've armed ourselves with the ability to critically evaluate platforms and tools, ensuring they meet the specific demands of RWD while considering the broader business implications of these choices. Understanding how to navigate these decisions empowers teams to build scalable, future-proof digital experiences that adapt to user needs and technological advancements.

We also delved into the art of managing vendor relationships effectively, a skill that enhances collaboration and project success. Recognizing the importance of these partnerships allows for a more seamless integration of services and technologies, driving the innovative potential of our RWD endeavors. Additionally, assessing platform flexibility and future scalability ensures that our digital assets remain relevant and engaging as the digital landscape evolves.

The knowledge and strategies you've gained in this chapter are invaluable. They not only streamline the development process but also position our projects to thrive in a competitive digital environment. By making informed decisions on platforms and vendors, we lay a solid foundation for RWDs that are not just visually appealing but also robust, user-friendly, and adaptable.

As we move forward to the next chapter, we'll build on these foundational decisions to further refine our approach to RWD. The next chapter will guide us in creating cohesive, immersive experiences that captivate users, regardless of how they access our digital content. The skills and insights from vendor and platform decisions will serve as a cornerstone for these advanced strategies in user engagement.

13

Engaging Users Across Touchpoints

In the digital realm, engaging users effectively across multiple touchpoints is not just a strategic advantage — it's also a necessity. This chapter dives into the nuances of creating seamless user experiences in a world where interactions span a multitude of devices, from smartphones and desktops to tablets and wearables. This chapter is dedicated to equipping you with the knowledge to understand and leverage the complexities of user engagement in the context of **responsive web design (RWD)**.

The topics covered in this chapter are as follows:

- Multi-touchpoint engagement in RWD
- Designing for touch, gesture, and beyond
- Consistency in multi-modal interactions
- Adapting to evolving user interaction trends

Multi-touchpoint engagement in RWD

In today's digital ecosystem, users interact with content across a multitude of devices and platforms, from smartphones and tablets to laptops, smartwatches, and beyond. This multi-touchpoint environment presents both challenges and opportunities for RWD. Understanding this ecosystem is crucial for designers and developers aiming to create seamless, engaging user experiences.

To grasp the intricacies of the multi-touchpoint ecosystem and its implications for RWD, let's take a look at the following concepts:

- **Understanding the nature of multi-touchpoint interactions**: Users no longer access digital content through a single device. A person might start reading an article on their smartphone during their commute, continue on a desktop at work, and finish on a tablet at home. Each of these touchpoints offers different user experiences, due to varying screen sizes, capabilities, and contexts of use. The challenge for RWD is to ensure consistency and fluidity across all these platforms.

- **Identifying the key touchpoints**: The first step in navigating the multi-touchpoint ecosystem is identifying the key touchpoints for your audience. Analytics can reveal which devices your audience uses, while user research can offer deeper insights into how, when, and why different devices are used. This understanding allows you to design strategies that cater to specific user needs and behaviors at each touchpoint.

- **Maintaining consistent yet flexible design**: RWD must balance consistency with flexibility. While visual identity and core functionality should remain consistent across devices, the design and interactions may need to adapt to the specific capabilities and use cases of each touchpoint. For example, touch gestures on mobile devices, hover states on desktops, and voice commands on smart assistants all require different design considerations.

- **Optimizing content for each touchpoint**: Content strategy plays a pivotal role in the multi-touchpoint ecosystem. Not all content needs to be available on all devices, but what is presented should be optimized for the context of each touchpoint. This might mean shorter, more actionable content on mobile devices and more in-depth, immersive experiences on larger screens.

- **Exploring the role of technology**: Emerging technologies such as **Progressive Web Apps (PWAs)** and **Adaptive Web Design (AWD)** are critical in addressing the challenges of multi-touchpoint engagement. PWAs, for example, can offer app-like experiences across all touchpoints, while AWD can serve different layouts or content based on the device being used.

- **Conducting continuous testing and iteration**: Engaging users across multiple touchpoints requires ongoing testing and iteration. Responsive designs should be regularly tested on a range of devices to ensure that they meet user needs and expectations. This includes not just technical performance but also user satisfaction and engagement metrics.

Understanding the multi-touchpoint ecosystem is about recognizing the diversity of user interactions in today's digital landscape and crafting RWD solutions that are flexible, consistent, and user-centered. By closely aligning design and content strategies with the realities of how users engage with digital content across devices, businesses can create more engaging, effective, and seamless online experiences.

- **Implementing RWD testing**: Identifying key user touchpoints across the diverse landscape of devices and platforms is a crucial step in optimizing RWD for multi-touchpoint engagement. Each touchpoint represents an opportunity to engage with the user, and understanding when and how users interact with your content can inform design decisions that enhance user experience.

Here's a guide to identifying these critical touchpoints and leveraging them to create a cohesive and engaging RWD strategy.

- **Analyzing user behavior and device usage**: Start by gathering data on how users interact with your site across different devices. Analytics tools can provide insights into device usage patterns, including the types of devices users access your site from, peak usage times, and the most frequently accessed content. This data helps pinpoint critical touchpoints where your RWD strategy should focus.

- **Mapping the user journey**: Mapping out the user journey offers a comprehensive view of the various touchpoints users encounter as they interact with your brand across devices. Include all potential interactions, from initial discovery via search engines on a smartphone, to browsing on a tablet, to completing transactions on a desktop. Understanding these pathways can highlight opportunities to improve the RWD experience at each stage.

- **Considering the context of use**: The context in which a device is used significantly affects user expectations and behavior. Smartphones are often used on the go, requiring quick, easily digestible information and simple navigation. Desktops, conversely, are typically used in more stable environments, allowing for deeper engagement with content. Tablets often blend these contexts, serving both as tools for leisurely browsing and productivity. Tailor your RWD to meet these contextual needs at each touchpoint.

- **Focusing on critical interactions**: Within the user journey, identify the interactions that are most critical to achieving your site's goals. This might include actions such as making a purchase, signing up for a newsletter, or contacting customer service. Ensure that these critical interactions are as seamless and intuitive as possible across all devices to encourage conversion and retention.

- **Leveraging qualitative feedback**: Analytics provide quantitative insights into device usage, but qualitative feedback from users offers valuable context about their experiences and preferences. Use surveys, user testing, and feedback forms to gather direct input from users about their cross-device experiences. This feedback can reveal pain points and preferences not immediately apparent in analytics data.

- **Regularly reassessing and adapting key user touchpoints**: The landscape of devices and user behavior is constantly evolving. Regularly reassess your key user touchpoints to ensure your RWD strategy remains aligned with current trends. Stay adaptable, being ready to refine your approach as new devices emerge and user preferences shift.

By meticulously identifying and understanding key user touchpoints, you can craft an RWD strategy that truly resonates with your audience, regardless of how or where they choose to engage with your content. This user-centered approach not only enhances the immediate experience but also builds a foundation for long-term engagement and loyalty.

Now, let's explore how to go about designing for touch-friendly interfaces.

Designing for touch, gesture, and beyond

Designing for touch-friendly interfaces is pivotal in creating responsive websites that are accessible and engaging across a wide range of devices. As touchscreens become more prevalent, from smartphones to tablets and even desktop monitors, the importance of implementing touch-friendly design principles cannot be overstated. Here are some key principles to ensure that your RWD caters effectively to touch interactions:

- One of the fundamental aspects of touch-friendly design is creating tap targets — such as buttons, links, and form fields — that are easy to interact with using fingers. A touch target that's too small or too close to other targets can lead to frustrating user experiences and accidental actions. As a general rule, aim for tap targets that are at least 48 pixels in height and width, with ample spacing around them to prevent mis-taps.

- When designing for mobile devices, consider how users hold their phones and primarily use their thumbs to interact. The most comfortable area for touch interaction is the center of the screen, which can be easily reached by the thumb without adjusting the grip. Place key actions and navigation elements within this thumb-friendly zone to enhance usability.

- Touchscreens don't have the hover state available to mouse-driven interfaces, which means designers need to rethink how to present interactive elements and additional information. Simplify interactions by making actions direct and results immediate, and consider alternative ways to reveal additional content, such as through tapping or a simple gesture.

- Gestures such as swiping, pinching, and double-tapping offer new dimensions of interaction beyond traditional clicks. When incorporating gestures into your design, use them intuitively and in ways that users expect — for example, swiping to scroll through a carousel or pinching to zoom in on a photo. Always provide visual cues or instructions for gesture interactions, and ensure that essential functions are not solely reliant on gestures to be accessible to all users.

- Filling out forms can be cumbersome on touch devices, especially if the form fields are not optimized for touch input. Make forms as simple as possible, reduce the number of fields, and use appropriate input types for different data (e.g., a numeric keypad for phone numbers). Providing an ample tapping area for checkboxes, radio buttons, and drop-down menus enhances the form-filling experience.

- Visual feedback is crucial in touch interfaces to confirm that an action has been recognized. This could be a subtle color change, animation, or a sound cue when users tap a button or complete an interaction. Immediate feedback helps users understand that their input has been successfully received and what to expect next.

- Finally, the best way to ensure your design is truly touch-friendly is to test it on actual touch devices. This allows you to experience your site as users would, identifying any areas where touch interactions might be less than optimal. Use a variety of devices with different screen sizes and operating systems to get a comprehensive understanding of the touch experience across your site.

Incorporating these principles of touch-friendly design into your RWD projects ensures that your websites are not only accessible but also delightful to interact with across all touch-enabled devices, contributing to a positive and seamless user experience.

Next, let's explore some gesture-based activities.

Gesture-based navigation and interactions

As interactive technology evolves, gesture-based navigation and interactions become increasingly integral in providing intuitive and engaging user experiences, especially in RWD. Gestures such as swiping, pinching, and tapping offer users a more natural and fluid way to navigate and interact with digital content across various devices. Here's how to effectively incorporate gesture-based navigation into your RWD projects:

- Start by understanding the common gestures used across different devices and platforms. Swiping is often used to navigate through a series of items, such as photos in a gallery. Pinching can zoom in or out on a piece of content while tapping typically selects or activates an item. Each gesture should feel intuitive to the user and align with the action it initiates.

- Gesture-based interactions should feel natural to the user. This means considering the physicality of the gesture and its context within the interface. For example, a swipe gesture to delete an item should be accompanied by visual feedback that confirms the action, such as an item moving off-screen. Ensure that gestures are consistent with users' expectations, based on their experiences with other apps and websites.

- Because gestures are invisible until performed, it's crucial to provide clear visual cues that hint at their availability. This can include icons, such as arrows for swipeable carousels, or instructional overlays for more complex gestures. Visual cues guide users on how to interact with your site, reducing confusion and enhancing the overall experience.

- While gestures can enhance navigation and interaction, overloading your site with too many gesture-based actions can confuse users and lead to accidental inputs. Stick to widely understood gestures and use them sparingly, ensuring that they complement rather than complicate the user experience.

- Not all users can perform all gestures, so it's vital to provide alternative navigation options. This might include traditional click-based navigation or keyboard shortcuts. Ensuring your site remains accessible to users with disabilities or those who prefer other forms of interaction is crucial in inclusive design.

- Testing is essential to ensure that gesture-based interactions work as intended across different devices and screen sizes. This includes not only technical performance but also user acceptance testing to gather feedback on how intuitive and satisfying users find the gestural interactions. Test a variety of touch-enabled devices to cover a broad spectrum of user scenarios.

- For less common gestures or unique interactions designed for your site, consider incorporating brief tutorials or tooltips that educate users on how to perform these gestures. User education can be particularly important for complex or novel interactions that users may not be familiar with.

Incorporating gesture-based navigation and interactions into your RWD strategy can significantly enhance the usability and delight of your site across touch-enabled devices. By designing intuitive gestural interactions, providing clear visual cues, ensuring accessibility, and conducting thorough testing, you can create a more dynamic and engaging user experience that leverages the full potential of modern touch technology.

Now, let's explore more than just touch.

Beyond touch — voice, VR, and AR in RWD

The realm of RWD is rapidly expanding beyond traditional touch interactions to incorporate voice commands, VR, and AR. These technologies offer new dimensions of engagement and present unique challenges and opportunities for designers. Understanding how to integrate voice, VR, and AR into RWD can significantly enhance the user experience, making digital content more accessible, immersive, and interactive.

Voice commands — enhancing accessibility and convenience

Voice interaction is becoming increasingly prevalent, thanks to the rise of smart speakers and voice assistants. Integrating voice commands into RWD allows users to navigate, search, and interact with web content hands-free, offering an alternative for users with disabilities or those seeking convenience. To incorporate voice effectively, do the following:

- Design **voice user interfaces** (**VUIs**) that understand natural language commands and queries

- Provide auditory feedback and confirmations to guide users through voice interactions

- Ensure content is structured semantically, making it easier for voice assistants to interpret and present information accurately

That wraps up the voice command list. Let's dive into the VR list.

Virtual reality — immersive web experiences

VR has the potential to transform web browsing into an immersive experience, allowing users to explore digital environments in a three-dimensional space. While primarily associated with gaming and entertainment, VR can also enhance educational content, virtual tours, and online shopping. Integrating VR into RWD requires the following:

- Designing for a 360-degree environment, considering user navigation and interaction within a virtual space

- Optimizing content for VR devices, ensuring fast load times and smooth performance
- Providing fallback options for users without VR headsets, allowing them to still access content in a traditional two-dimensional format

That wraps up VR. Let's dive into the AR list.

Augmented reality — blending digital and physical worlds

AR overlays digital content in the real world, offering unique opportunities for interactive and contextual web experiences. From trying on clothes virtually to visualizing furniture in a room, AR can bridge the gap between online content and physical reality. To effectively integrate AR into RWD, do the following:

- Develop AR experiences that add value and enhance user engagement without being gimmicky
- Ensure that AR content is responsive, adapting to various device cameras and screen sizes
- Consider user privacy and safety, particularly when using AR in public or shared spaces

That wraps up the AR list. Let's explore some other design considerations.

Design considerations for voice, VR, and AR in RWD

Integrating these technologies into RWD involves several key considerations:

- **User consent and control**: Always allow users to opt in to voice, VR, and AR experiences, providing clear options to control or disable these interactions.
- **Accessibility**: Ensure that these advanced interactions do not exclude users with disabilities. Offer alternative ways to access the same content and functionalities.
- **Performance optimization**: Voice, VR, and AR content can be resource-intensive. Optimize for performance to ensure a smooth user experience across all devices.
- **Testing and feedback**: Regular user testing is crucial to refine voice, VR, and AR experiences, ensuring that they meet user needs and expectations.

By embracing voice, VR, and AR technologies, RWD can offer more diverse and engaging experiences that cater to the evolving ways users interact with digital content. These technologies represent the frontier of web design, challenging designers to think beyond the screen and consider a future where digital experiences are more integrated into our physical world.

Now, let's dive into how to create consistency with multi-modal interactions.

Consistency in multi-modal interactions

Ensuring consistency in visual design across different modes of interaction is pivotal for creating a seamless user experience in RWD. As users shift between devices and interaction methods — from touch and mouse to voice commands, AR, and VR — maintaining a harmonious visual design ensures that a brand's identity remains intact and recognizable. This coherence supports user familiarity, enhances usability, and strengthens brand perception.

Here's how to achieve visual harmony across diverse interaction modes:

- Start by establishing a comprehensive design language that encapsulates your brand's visual identity. This includes defining color schemes, typography, iconography, and imagery that resonate with your brand's essence. A well-defined design language serves as the foundation to create consistent experiences across all platforms and modes of interaction.

- While the core elements of your design language should remain consistent, they also need to be adaptable to different interaction contexts. For instance, the way your design translates to a VR environment might differ in its spatial implementation compared to a flat touchscreen, but key elements such as color and typography should be recognizable across these modes. Consider creating flexible design systems that allow for this adaptability without losing the essence of your brand.

- Beyond visual elements, ensure that navigation structures and interaction patterns are consistent across modalities. For example, if a swipe gesture performs a specific action in your mobile app, a similar action in AR or VR environments (such as waving a hand) should elicit a comparable response. Consistency in interaction patterns reinforces user learning and reduces cognitive load.

- Symbols and icons play a crucial role in guiding users across different interaction modes. Utilize universally recognized icons, and ensure that they are consistent in appearance and function across all modes. This not only aids in immediate recognition but also helps bridge the gap between different interaction methods.

- Consistency in visual design also entails ensuring that your content is accessible across all modes of interaction. Use high-contrast color schemes, legible typography, and clear icons that are easily decipherable, regardless of the interaction mode. Including alternative text descriptions and voice command feedback ensures that users who rely on assistive technologies can also enjoy a consistent experience.

- Testing your design across various modes of interaction is crucial for identifying inconsistencies and areas for improvement. Gather feedback from users who interact with your design through different touchpoints to understand how they perceive your brand across these modalities. This feedback is invaluable for refining your visual design to better serve a multi-modal interaction landscape.

Harmonizing visual design across different modes of interaction requires thoughtful consideration of how core design elements can adapt while maintaining a consistent brand identity. By focusing on adaptability, consistent navigation and interaction patterns, universal symbols, accessibility, and thorough testing, designers can create cohesive and engaging experiences that resonate with users across any touchpoint or device.

Next, let's explore how we can synchronize behavior across platforms.

Synchronizing functional behavior across platforms

Consistency in functional behavior across platforms ensures that users have a seamless experience, regardless of how or where they interact with your digital presence. As users switch between devices and interaction modes — from desktops to smartphones, and from touch to voice, or even AR — the way your website or application responds should remain predictable and reliable. Here are some strategies to synchronize functional behavior across platforms:

- Start by mapping out the core functionalities of your website or application, identifying how each feature operates across different platforms and modes of interaction. This framework should detail the expected behavior for common actions, such as navigating menus, submitting forms, or performing searches, ensuring that these actions are intuitive and consistent regardless of the user's access point.

- Adopting a mobile-first approach in RWD not only prioritizes performance and usability on smaller screens but also ensures that functional behaviors are designed to be adaptable from the outset. By starting with the most constrained environment, you can ensure that functionalities developed for mobile platforms are inherently flexible, making it easier to scale up and adapt them for larger screens and additional interaction modes.

- Progressive enhancement involves starting with a basic level of user experience that all browsers and devices can access and then adding more sophisticated layers of interaction that enhance the experience where possible. This approach ensures that essential functionalities remain consistent across platforms, while still taking advantage of more advanced capabilities on devices that support them.

- Use device and mode detection techniques to tailor functional behaviors to the capabilities of the user's device or preferred mode of interaction. For instance, a website could adjust its navigation structure based on whether it's being accessed via touch, mouse, or voice command. However, it's crucial to ensure that these adjustments are not obstructive and that alternative access methods are always available, maintaining functional consistency.

- Comprehensive testing across a wide range of devices and interaction modes is critical for identifying discrepancies in functional behavior. This should include not only technical testing to ensure compatibility but also user testing to gauge how intuitively functionalities translate across different platforms. Pay special attention to transition points where users might switch between devices or modes, ensuring that these shifts are as seamless as possible.

- Consistency in functional behavior is closely tied to accessibility and usability. Ensure that all users, including those with disabilities, can access and utilize the core functionalities of your site or app in similar ways, regardless of the platform or interaction mode. This might involve providing keyboard navigability, voice command functionality, and clear visual feedback across all devices.

- The digital landscape is continuously evolving, with new devices, platforms, and interaction modes emerging regularly. Maintaining an agile approach to design and development allows you to adapt and update your functional behaviors in response to these changes, ensuring ongoing consistency across platforms.

Synchronizing functional behavior across platforms is crucial for a cohesive user experience in multi-modal digital environments. By establishing a unified functional framework, adopting mobile-first and progressive enhancement strategies, and ensuring thorough testing and accessibility, you can create a responsive design that offers reliable and consistent functionality for all users, regardless of how they choose to engage.

Next, let's investigate universal accessibility in multi-modal design.

Universal accessibility in multi-modal design

Universal accessibility in multi-modal design is not just a goal — it's also a necessity to create inclusive digital experiences that cater to all users, including those with disabilities. As RWD evolves to incorporate a variety of interaction methods, including touch, voice commands, AR, and VR, ensuring accessibility across these modes becomes increasingly complex yet imperative. Here's how to ensure universal accessibility in multi-modal designs:

- Accessibility should encompass the full range of human abilities, considering visual, auditory, motor, and cognitive impairments. This broad approach ensures that your design is accessible regardless of how users interact with your content — be it through touch, speech, gestures, or other modes. Recognize that each user's experience is unique, and design for flexibility and adaptability.

- At the core of accessible web design is compatibility with keyboard navigation and screen readers, which are essential for many users with disabilities. Ensure that all interactive elements are reachable and operable through keyboard inputs and that content is structured logically for screen reader users. This foundational level of accessibility benefits users across all modalities.

- Voice interaction provides an alternative access point for users who may find traditional navigation methods challenging. Integrate voice commands into your RWD, allowing users to navigate, select, and interact with content through spoken commands. Ensure that voice interfaces are designed to understand and respond to natural language, offering feedback and guidance to users.

- ARIA landmarks and roles help define the structure of web content, making it easier for assistive technologies to interpret and navigate. When designing for multi-modal interactions, use ARIA landmarks to denote navigation menus, main content, forms, and other significant areas of your page. This aids users in quickly understanding and moving through your content, regardless of the interaction mode they're using.

- While touch, AR, and VR offer immersive ways to engage with digital content, not all users can fully utilize these modalities. Provide alternative means of access for content and functionalities available through advanced interactions. For instance, if a product view is enhanced through AR, it should also offer high-quality images and descriptions for users who cannot engage with AR content.

- User testing is critical to identifying accessibility issues within your multi-modal design. Include a diverse group of users in your testing phases, especially those who use assistive technologies or have varying abilities. Their insights can highlight unseen barriers and provide valuable feedback on how to improve universal accessibility across your design.

- The landscape of technology and accessibility standards is ever-evolving. Stay informed about the latest developments in assistive technologies, interaction modes, and web accessibility guidelines. Regularly update your knowledge and your designs to reflect these advancements, ensuring that your RWD remains accessible to everyone.

Universal accessibility in multi-modal design ensures that all users, regardless of their abilities or the technologies they use, can have a rich, engaging, and seamless experience with your digital content. By integrating these principles into your RWD strategy, you affirm your commitment to inclusivity and enhance the usability and reach of your digital presence.

Let's break down how we can adapt to evolving user interaction trends.

Adapting to evolving user interaction trends

In the rapidly evolving landscape of digital interaction, staying abreast of emerging trends is crucial for RWD professionals aiming to create engaging and future-proof user experiences. Monitoring and adapting to these trends not only enhances the accessibility and usability of web content across devices but also ensures that designs remain relevant in the face of technological advancements.

Here's how to effectively monitor and adapt to evolving user interaction trends:

- Dedicate time to reading industry publications, research papers, and trend reports from leading tech companies and design institutions. These sources often provide early insights into emerging technologies and user interaction patterns, offering a glimpse into the future of digital design and how it could impact RWD.

- Engaging with online forums, social media groups, and professional networks focused on web design and technology allows you to tap into collective knowledge and experiences. These communities can be valuable resources for discovering new interaction trends, sharing insights, and discussing the potential implications for responsive design practices.

- Technology and design conferences, workshops, and webinars are excellent opportunities to learn from thought leaders and innovators in the field. These events often showcase the latest trends in user interaction, including demonstrations of cutting-edge technologies such as AR, VR, and voice interfaces, providing inspiration and practical knowledge for RWD professionals.

- Hands-on experimentation with emerging technologies can offer invaluable insights into their potential applications in RWD. Whether it's building prototypes using new frameworks or testing out new hardware, direct experience helps you understand the challenges and opportunities these technologies present for user interaction design.

- Incorporate emerging interaction trends into your design prototypes, and conduct user testing to gather feedback on their effectiveness and usability. Observing real users as they interact with new features or technologies can reveal unforeseen issues and opportunities for improvement, guiding the refinement of your responsive designs.

- Leverage analytics tools to monitor how users interact with your site or application, paying special attention to emerging patterns that could indicate shifting preferences or behaviors. Data on device usage, navigation paths, and engagement metrics can provide clues to changing user expectations and help you anticipate the need for design adjustments.

- In the fast-paced world of technology and design, fostering a culture of continuous learning within your team is essential. Encourage ongoing education, share new findings and insights regularly, and create an environment where experimentation and innovation are valued. This culture ensures your team remains agile and capable of adapting to new user interaction trends as they arise.

By actively monitoring emerging interaction trends and integrating this knowledge into your RWD strategies, you can create experiences that not only meet current user expectations but are also positioned to evolve with future advancements. This proactive approach to design ensures that your digital content remains engaging, accessible, and relevant, no matter how user interaction trends may shift in the future.

Next, let's evaluate new trends.

Evaluating new trends for RWD implementation

As the digital landscape continuously evolves, staying ahead involves not just monitoring emerging user interaction trends but also meticulously evaluating their relevance and potential impact on RWD. Implementing new trends requires a strategic approach to ensure that they enhance the user experience without detracting from the core objectives of accessibility, usability, and performance.

Here are the steps to evaluate new trends for RWD implementation:

1. **Assess relevance to the target audience**: Before embracing any new trend, assess its relevance to your target audience. Conduct user research to understand their preferences, technology usage patterns, and specific needs. A trend that resonates well with your audience can significantly enhance engagement, while one that doesn't might waste resources without adding value.

2. **Check compatibility with design goals**: Evaluate how well the trend aligns with your project's design goals and overall strategy. Consider whether it supports your objectives, such as improving navigation, increasing engagement, or enhancing accessibility. The trend should complement your existing design framework rather than complicate or contradict it.

3. **Analyze the feasibility and technical requirements**: Analyze the feasibility of integrating the trend into your RWD projects, considering the technical requirements, resource availability, and potential development challenges. Some trends might require advanced technologies or substantial modifications to existing structures, making them less practical for certain projects.

4. **Assess the potential impact on performance**: Performance is a critical aspect of RWD, affecting everything from user satisfaction to search engine rankings. Assess the potential impact of implementing the trend on your website or application's performance. Ensure that any new feature is optimized for speed and efficiency across all devices and connection types.

5. **Take accessibility considerations into account**: Any new trend or feature must be accessible to all users, including those with disabilities. Evaluate how the trend can be implemented in an accessible manner, ensuring it adheres to web accessibility guidelines and standards. Incorporating trends that enhance rather than hinder accessibility can broaden your audience and improve overall user experiences.

6. **Conduct testing and listen to user feedback**: Before fully committing to a trend, conduct prototype testing with real users to gather feedback on its functionality and appeal. This can provide valuable insights into how the trend is perceived and its actual impact on the user experience. User feedback can guide further refinements or indicate whether a trend should be adopted at all.

7. **Concentrate on monitoring and iteration**: After implementing a new trend, continuously monitor its performance and user interaction metrics. Be prepared to iterate based on real-world usage and feedback. The digital environment is dynamic, and user preferences can shift quickly; being agile and responsive to these changes is key to maintaining an effective RWD.

Evaluating new trends for RWD implementation requires a balanced approach that considers the needs and preferences of your audience, the alignment with your design goals, and the practical aspects of integration. By thoughtfully assessing each trend's potential impact, you can ensure that your RWD initiatives remain innovative and user-centric, providing meaningful and engaging experiences across all touchpoints.

Next, let's explore how we can use agile integration for other things.

Agile integration of new interaction modes

Incorporating new interaction modes into RWD necessitates an agile approach to integration — one that allows for the rapid iteration and deployment of features that align with evolving user expectations and technological advancements. As user interaction trends shift toward more dynamic and diverse modes, such as voice commands, AR, and gesture-based navigation, adopting an agile methodology ensures that RWD remains both current and forward-thinking.

Here's how to agilely integrate new interaction modes into RWD:

- Modular design principles are at the heart of agile RWD. By building your website or application with interchangeable components, you can easily add, remove, or update interaction modes without overhauling the entire system. This modularity allows for the seamless introduction of new features and ensures that your design can evolve with technological trends and user preferences.

- Agile integration thrives on collaboration across disciplines — designers, developers, content creators, and **user experience** (**UX**) specialists must work closely to explore and implement new interaction modes. Regular brainstorming sessions and collaborative workshops can spark innovative ideas and ensure that all aspects of the user experience are considered in the design process.

- Rapid prototyping is essential to test and refine new interaction modes. Tools that enable the quick creation and iteration of design concepts allow teams to explore various implementations and assess their usability, functionality, and impact on the overall UX. Prototypes should be tested across a range of devices and user scenarios to identify potential issues and opportunities for enhancement.

- User feedback is invaluable in the agile integration process. Conducting ongoing user testing sessions with prototypes and beta releases allows you to gather direct input on how new interaction modes are received by your target audience. This feedback informs further iterations, helping to refine the implementation and ensure that it meets user needs and expectations.

- Keeping abreast of the latest trends in technology and user interaction is crucial for agile RWD. Attend conferences, follow industry leaders on social media, and participate in professional forums to stay informed about emerging interaction modes and their potential applications. An open-minded and adaptable approach ensures that your designs remain at the forefront of digital innovation.

- Use web analytics tools to monitor how users engage with new interaction modes once they're implemented. Analyzing behavior patterns, engagement metrics, and conversion rates provides insights into the effectiveness of these features. This data-driven approach enables continuous optimization, ensuring that new interaction modes enhance rather than hinder the UX.

- As new interaction modes are integrated, accessibility should remain a priority. Ensure that all users, including those with disabilities, can benefit from the enhanced functionality. This may involve providing alternative navigation options or ensuring that voice and gesture controls are compatible with assistive technologies.

Agile integration of new interaction modes into RWD requires a structured yet flexible approach, emphasizing modular design, cross-disciplinary collaboration, rapid prototyping, and user-centered testing. By staying informed about emerging trends and leveraging analytics for insights, you can ensure that your responsive designs continue to meet and exceed user expectations in an ever-evolving digital landscape.

Summary

As we wrap up *Chapter 13*, it's evident that the landscape of user engagement is both vast and varied. Throughout this journey, we've uncovered the critical importance of crafting consistent and compelling experiences across all devices, recognizing that each touchpoint plays a pivotal role in the overall perception of your digital presence. We've delved into understanding user behaviors and preferences across different devices, equipping ourselves with the insight to tailor experiences that not only meet but also exceed user expectations.

The strategies discussed to maintain engagement through transitions between devices are invaluable. They serve as a roadmap to create fluid, intuitive, and accessible interactions, regardless of the medium. By implementing these strategies, we ensure that our digital solutions are not just seen but also felt, fostering a deeper connection with our audience.

Analyzing touchpoint effectiveness has underscored the necessity of a data-driven approach to design and development. Adjusting strategies based on user feedback and interaction data ensures that our efforts are not static but evolve with the needs and behaviors of our users. This iterative process is fundamental to staying relevant in a digital space that is perpetually in flux.

The skills and lessons gleaned from this chapter are instrumental in navigating the complex web of user interactions in today's multi-device world. They empower us to create more engaging, effective, and memorable digital experiences that resonate across all touchpoints.

Moving forward to the next chapter, we build upon the foundation laid in engaging users across touchpoints to explore how we can ensure the longevity and relevance of our digital strategies. As the digital landscape continues to evolve at a rapid pace, understanding how to anticipate and adapt to future trends is crucial. This next chapter will guide us in solidifying our responsive design efforts to not only meet current demands but also embrace the challenges and opportunities of tomorrow's digital ecosystem.

14

Future-Proofing Responsive Strategies

As we venture deeper into the digital age, the need for RWD strategies that not only address current requirements but also anticipate future trends becomes increasingly paramount. This chapter is a comprehensive guide that's designed to equip you with the foresight and adaptability needed to navigate the ever-evolving landscape of web design and user interaction. This chapter is dedicated to understanding how to stay ahead in the rapidly changing digital environment, ensuring that RWD efforts remain effective, relevant, and user-centric for years to come.

The digital world is in a state of constant flux, with emerging trends and technologies reshaping the way users interact with online content. This chapter starts by emphasizing the importance of keeping abreast of these changes, providing insights into how to monitor and evaluate new developments in RWD. It will guide you through the process of adapting to changing user behaviors and device preferences, highlighting the need for RWD strategies that are not only responsive but also anticipatory.

Moreover, planning for the long-term evolution of RWD strategies is crucial. This chapter offers practical advice on building flexible, scalable web designs that can easily integrate future advancements. It delves into the challenges and opportunities presented by new technologies such as AI, VR, and voice interfaces, exploring how they can be seamlessly incorporated into existing frameworks without disrupting the user experience.

This chapter is more than just a guide to keeping up with the times; it's about setting the pace and establishing practices that ensure sustained success and relevance in the digital space.

The following topics will be covered in this chapter:

- Predicting the evolution of web interactions
- The role of AI and AR in responsive futures
- Ensuring longevity in design choices
- Staying agile in a rapidly evolving digital space
- Agile methodologies in RWD

By the end of this chapter, you will be equipped with the knowledge and tools needed to adapt your RWD efforts to the not-so-distant future, where the only constant is change.

Predicting the evolution of web interactions

Reflecting on the historical trends of web interactions offers us a window into the potential future of digital design and user engagement. This journey, from the static pages of the early internet to the dynamic and interactive experiences we see today, highlights a trajectory of increasing personalization, interactivity, and immersive experiences. As we look forward, anticipating the next wave of innovation becomes a critical task for designers and strategists aiming to future-proof their responsive strategies. Looking ahead, the evolution of web interactions is poised to become even more integrated into our physical world. The **Internet of Things** (**IoT**) promises a future where web interactions extend beyond screens into our everyday objects, making responsive design more critical than ever. Voice interfaces and natural language processing will likely become standard features, making web access more intuitive and barrier-free.

Predictive technologies will further personalize user experiences, with systems anticipating needs and tailoring content in real time. Furthermore, the advancement of VR and the continued growth of AR could revolutionize how we perceive web interfaces, with spatial web design becoming a key consideration.

Future-proofing responsive strategy means staying adaptable, open to adopting new technologies, and continually focusing on the user's evolving needs. It involves creating flexible design systems that can accommodate not just today's devices but also tomorrow's innovations. Designers and strategists must remain vigilant, always ready to learn and apply the lessons of the past and present to the unseen challenges and opportunities of the future.

In conclusion, understanding historical trends and projecting future developments requires a balance of creativity, technical acumen, and strategic foresight. As we move forward, the capacity to anticipate and innovate in response to changing web interactions will define the success of responsive web strategies, ensuring they remain relevant, engaging, and accessible for all users, regardless of how technology evolves.

Let's explore the impact of emerging technologies.

The impact of emerging technologies

Emerging technologies are catalysts for transformation in the digital world, significantly impacting the evolution of web interactions. These advancements promise to redefine the boundaries of what's possible online, offering new opportunities for immersive, interactive experiences that could reshape the landscape of RWD. Understanding the implications of these technologies is pivotal for designers and developers aiming to future-proof their strategies and create forward-thinking, adaptable digital environments.

AI and ML are at the forefront of this technological revolution, driving personalized experiences to new heights. AI's ability to analyze user data and predict behaviors enables websites to offer highly tailored content, recommendations, and interactions. For RWD, this means designs that not only adapt to the device but also the user's unique preferences and needs, creating a more intuitive and engaging web experience.

AR and VR technologies are expanding the web's spatial dimensions, allowing users to engage with content in three-dimensional, immersive environments. AR overlays digital information onto the physical world through the user's device, enhancing the shopping experience with virtual try-ons and interactive product demos. VR, meanwhile, creates entirely digital environments for users to explore, such as games that let you explore fantasy worlds. For RWD, integrating these technologies requires a thoughtful design that considers the varied capabilities of devices, ensuring accessibility while pushing the boundaries of what web experiences can be.

IoT extends web interactions beyond traditional computing devices, embedding connectivity in everyday objects. This interconnectedness offers new touchpoints for engaging with users, from smart home devices to wearables. RWD must evolve to consider these varied contexts, designing experiences that are fluid and coherent across an expanding ecosystem of connected devices.

The rollout of 5G networks is set to supercharge the web with faster, more reliable internet access. This leap in connectivity opens the door for more data-intensive applications and services, from high-definition streaming to real-time interactions without lag. For RWD, 5G offers the potential to design more complex, dynamic web experiences without compromising speed or performance, even on mobile devices.

Blockchain technology, while primarily associated with cryptocurrencies, offers intriguing possibilities for secure, decentralized web interactions. Its potential for creating trust and transparency online could revolutionize how we conduct transactions, share data, and authenticate information. As this technology matures, RWD strategies may need to adapt to incorporate these new standards of security and user control.

To leverage these emerging technologies effectively within RWD, designers, and developers must embrace a mindset of continuous learning and experimentation. Staying informed about technological advancements, participating in professional communities, and investing in ongoing training are essential for staying ahead of the curve. Additionally, adopting a flexible, modular approach to web design ensures that new technologies can be integrated seamlessly, allowing websites to evolve alongside these innovations.

In summary, the impact of emerging technologies on the evolution of web interactions is profound, offering exciting possibilities for creating more personalized, immersive, and interconnected web experiences. By understanding these technologies and their implications for RWD, professionals can ensure their digital strategies remain relevant and effective in the face of rapid technological change.

Now, let's learn about the importance of user behavior.

User behavior analysis and prediction models

In the rapidly evolving digital landscape, understanding and anticipating user behavior is paramount for creating RWDs that not only meet current needs but are also adaptable to future trends. User behavior analysis and prediction models are critical tools in this endeavor, enabling designers and developers to foresee changes in how users interact with digital content and interfaces. These insights are foundational for future-proofing responsive strategies, ensuring they remain effective and engaging as user expectations and technology evolve.

The first step in analyzing user behavior is gathering data. This involves tracking how users interact with a website across different devices, noting patterns such as click-through rates, navigation paths, time spent on pages, and bounce rates. Advanced analytics tools and heatmaps can offer a deeper understanding of user engagement, highlighting areas of interest and potential friction points.

With a robust dataset in hand, the next step is to develop models that can predict future behavior. ML algorithms can analyze historical data to identify trends and patterns, predicting how users are likely to behave under similar conditions in the future. These models can forecast shifts in user preferences, such as an increasing preference for voice search or a rising demand for AR experiences.

One of the most powerful applications of user behavior prediction is the personalization of web experiences. By anticipating individual user needs and preferences, websites can dynamically adjust content, layout, and interaction modes to suit each visitor. This could mean presenting a simplified navigation menu on mobile devices based on predicted user tasks or offering personalized content recommendations to enhance engagement.

Prediction models also facilitate scenario planning, allowing designers to test how changes in web interactions could affect user behavior. By simulating different future scenarios, from the adoption of new technologies to changes in user demographics, designers can assess potential impacts on the user experience and identify necessary adjustments to responsive strategies.

The digital environment is in constant flux, necessitating a continuous cycle of analysis, prediction, and adaptation. Incorporating user feedback and new data into prediction models ensures they remain accurate and relevant. This iterative process allows responsive designs to evolve in step with user behavior, ensuring they consistently meet or exceed expectations.

While user behavior analysis and prediction models are invaluable, they also present challenges. Privacy concerns and data protection regulations necessitate careful handling of user data. Additionally, the accuracy of predictions can vary, requiring ongoing refinement of models and strategies based on actual user feedback and behavior. This means that we need to be careful and selective in how we go about using user behavior analysis and prediction models.

User behavior analysis and prediction models are indispensable tools for future-proofing RWD. By understanding and anticipating how user interactions evolve, designers can create adaptable, user-centric experiences that stand the test of time. The key to success lies in leveraging data intelligently, respecting user privacy, and remaining agile in the face of changing digital trends.

From here, we must understand the role that AI and AR play in RWD.

The role of AI and AR in responsive futures

The integration of AI and AR into RWD marks a significant leap toward creating more intuitive, personalized user experiences. AI-driven personalization is revolutionizing how users interact with web content, making these interactions more relevant, engaging, and user-friendly. This approach not only enhances user satisfaction but also significantly boosts the effectiveness of web platforms in meeting diverse user needs.

AI — it's taking over

AI excels in analyzing vast amounts of data to identify patterns, preferences, and behaviors of individual users. By leveraging this capability, RWD can dynamically adapt not just to the size of the screen but also to the context and preferences of the user. For instance, an e-commerce site can display products based on a user's browsing history, purchase records, and even social media activity, all in real time and across different devices.

The power of AI extends to predicting user needs even before they explicitly express them, enabling a level of anticipatory design that was previously unattainable. Websites can suggest content, offer help through chatbots, or adjust navigation based on the user's predicted intent. This proactivity in design ensures that users find value at every touchpoint, improving overall satisfaction and engagement.

AI-driven RWD takes into account the uniqueness of each device and its usage context, delivering tailored experiences accordingly. On mobile devices, where screen space is limited, AI can prioritize content and functionalities based on the user's immediate needs and past interactions. On larger screens, it can afford to offer more detailed information and richer interactions, all while maintaining a coherent user experience across devices.

Incorporating AI into RWD is not without its challenges. It requires a robust infrastructure capable of processing and analyzing data in real time. Designers and developers must work closely to ensure that the AI's decisions align with the overall design philosophy and user experience goals. Privacy and ethical considerations are paramount as personalization relies on accessing and interpreting user data.

The future of AI in RWD promises even more sophisticated personalization and user experience enhancements. Technologies such as ML and natural language processing could enable websites to understand and respond to user queries with unprecedented accuracy and relevance. Moreover, as AI technologies become more accessible and integrated into web development tools, their adoption in RWD is set to increase, pushing the boundaries of what's possible in creating responsive, user-centric web environments.

AI-driven personalization represents a paradigm shift in how we approach RWD. By making websites more adaptable not just to devices but to individual users, AI is setting a new standard for personalized user experiences. For businesses and designers, the message is clear: embracing AI in RWD is no longer an option but a necessity for staying competitive and relevant in the digital age. As we move forward, the synergy between AI and RWD will continue to evolve, offering exciting possibilities for engaging users in more meaningful, personal ways.

Now, let's explore the use case for AR.

AR — bridging digital and physical worlds

AR stands at the forefront of merging the digital and physical realms, providing a transformative medium through which web interactions can be reimagined. In the context of RWD, AR offers an unparalleled opportunity to enrich user experiences by overlaying digital information onto the physical world. This integration not only enhances user engagement but also opens up new avenues for interaction that were previously confined to the realms of science fiction.

AR's potential to elevate user engagement is vast. Retailers, for example, can leverage AR to allow customers to visualize products in their own space before making a purchase, significantly reducing the gap between online shopping and the tangible, in-store experience. This direct interaction boosts confidence in purchase decisions and enhances user satisfaction, bridging the digital-physical divide.

Beyond retail, AR has profound implications for education and training. By superimposing educational content onto the real world, learners can interact with complex subjects in a hands-on manner, leading to improved comprehension and retention. For RWD, this means designing web platforms that can seamlessly integrate AR features, ensuring educational content is accessible and engaging across all devices.

Incorporating AR into RWD poses a set of challenges. The diversity of devices and varying degrees of technological capability necessitate careful consideration to ensure AR experiences are consistently high quality and accessible. Furthermore, designers must navigate the balance between innovative AR interactions and the usability principles that underpin effective RWD, ensuring that AR enhancements complement rather than complicate the user experience.

The development of AR-friendly web platforms is supported by a growing suite of tools and technologies. WebXR, for instance, is an API that enables the creation of VR and AR experiences directly in the browser, making it easier for web developers to incorporate AR into responsive designs without the need for specialized software or apps. This democratization of AR technology is crucial for its widespread adoption in RWD.

Looking ahead, the integration of AR into RWD is set to deepen, with emerging technologies making AR experiences more immersive and interactive. Advances in computer vision and AI will enable more sophisticated recognition of physical environments, allowing for more seamless and intuitive interactions between the user and the digital overlays. As 5G technology becomes more prevalent, the increased bandwidth and lower latency will further enhance the feasibility and fluidity of AR experiences on web platforms, making them an integral part of the user's digital journey.

AR's capacity to blend digital information with the physical world offers a promising horizon for enhancing RWDs. By effectively leveraging AR, designers and developers can unlock new dimensions of user interaction and engagement, making web experiences more immersive, personalized, and contextually relevant. As we navigate the evolving landscape of web technology, AR stands as a pivotal force in shaping the future of how we interact with the digital world, promising a future where the boundaries between digital and physical are increasingly blurred.

Now, let's explore the challenges and opportunities that AR and AI provide.

The challenges and opportunities of using AI and AR for RWD

As RWD continues to evolve, the integration of AI and AR presents both significant opportunities and notable challenges. These technologies offer the potential to dramatically enhance user experiences, making web interactions more intuitive, personalized, and immersive. However, their incorporation into RWD necessitates careful consideration of various technical, ethical, and design-related factors.

AI and AR can transform RWD by creating more dynamic and responsive interfaces that adapt to individual user preferences and context. AI's ability to analyze user behavior and predict needs can lead to more personalized web experiences, while AR can offer immersive experiences that blur the lines between the digital and the physical world. Together, they can significantly increase engagement, satisfaction, and conversion rates by offering richer, more interactive web experiences.

However, integrating AI and AR into RWD also introduces several technical challenges. Ensuring compatibility across a wide range of devices and browsers is paramount, as is optimizing performance to prevent slow loading times or lag, which can detract from the user experience. Moreover, developing AR experiences that function seamlessly in a web environment requires advanced skills and understanding of both AR technologies and RWD principles.

The use of AI in personalizing user experiences raises ethical considerations related to privacy and data protection. Designers and developers must navigate the fine line between personalization and intrusion, ensuring that user data is handled responsibly and transparently. Similarly, AR experiences must be designed with user safety in mind, avoiding scenarios that could lead to physical harm or disorientation.

The sophisticated nature of AI and AR technologies means that their successful integration into RWD demands a high level of expertise. Currently, there exists a skill gap in the workforce, with a shortage of professionals who possess both the technical skills to implement AI and AR and the design skills to incorporate them into responsive web environments. Bridging this gap is crucial for the future development of AI and AR-enhanced RWD.

Despite these challenges, the potential of AI and AR to revolutionize RWD is immense. By leveraging these technologies, designers, and developers can create web experiences that are not only more engaging and interactive but also more accessible and intuitive. The key to unlocking this potential lies in ongoing education, experimentation, and a commitment to ethical design practices.

AI and AR represent the cutting edge of what's possible in RWD, offering unparalleled opportunities to enrich user experiences. However, realizing these opportunities requires overcoming significant challenges, from technical hurdles to ethical dilemmas. As we look to the future, the successful integration of AI and AR into RWD will depend on our ability to navigate these challenges, harnessing the full potential of these technologies to create RWDs that are not only visually stunning but also deeply user-centric and ethically sound.

How can we ensure that our design choices are made to last?

Ensuring longevity in design choices

In the rapidly evolving digital landscape, creating a web design that withstands the test of time is a challenging task. The concept of timeless web design revolves around developing a system that remains functional, relevant, and aesthetically pleasing, despite changes in trends and technologies. Ensuring longevity in design choices means focusing on foundational principles that transcend the fleeting preferences of the digital era.

The core of timeless design lies in simplicity and clarity. A clean, uncluttered layout with a clear hierarchy not only enhances usability but also ensures that the design remains accessible and engaging across various devices and screen sizes. Simple designs are easier for users to navigate, making the website more adaptable to future changes in user interface patterns.

Content is the cornerstone of any effective web design. Timeless designs prioritize content delivery and readability, ensuring that users can easily find and engage with the information they need. This involves using legible fonts, effective contrast, and spacing that improves readability across all devices, thereby future-proofing the design against shifts in content consumption habits.

Timeless web designs employ responsive techniques that seamlessly adapt to any screen. This adaptability is crucial for longevity as it ensures that the website functions optimally across the ever-expanding spectrum of devices and resolutions. A flexible, responsive foundation minimizes the need for future redesigns, allowing the website to evolve with technological advancements without losing its core functionality.

Timeless designs are deeply rooted in a user-centric approach, focusing on creating a positive and engaging user experience. This involves understanding and anticipating user needs, behaviors, and preferences, and designing interfaces that are intuitive and easy to use. By prioritizing the user experience, designers can create web designs that remain relevant and effective over time.

Timeless web designs adhere to universal design principles, ensuring that the website is accessible to as wide an audience as possible, including users with disabilities. This includes adhering to **Web Content Accessibility Guidelines** (**WCAG**) and designing for inclusivity, thereby extending the website's relevance and usability across different user groups and devices.

While aesthetic trends come and go, timeless designs often opt for aesthetic neutrality, avoiding overly trendy elements that may quickly become outdated. This doesn't mean the design must be bland; rather, it should leverage classic visual elements that maintain their appeal over time, allowing for easy updates without comprehensive overhauls.

Finally, timeless web design is characterized by a commitment to quality in both aesthetics and functionality. High-quality images, typography, and interactive elements ensure the design remains appealing and functional long into the future. Quality also extends to the code base, with clean, well-documented code ensuring that the site is maintainable and scalable.

Ensuring longevity in web design requires a balanced approach that incorporates simplicity, flexibility, user-centricity, and a commitment to quality. By grounding design choices in these timeless principles, developers and designers can create websites that not only meet the demands of today's digital landscape but also adapt gracefully to the unknown challenges and opportunities of the future.

Let's explore what is meant by the terms *adaptable design systems* and *modular architecture*.

Adaptable design systems and modular architecture

Creating web designs that remain relevant and functional over time requires a foundational approach that embraces adaptability and modularity. Adaptable design systems and modular architecture stand at the forefront of this strategy, offering a technique that can evolve with technological advancements and changing user needs without necessitating complete overhauls.

Let's break down adaptable design systems first.

Exploring adaptable design systems

An adaptable design system is more than just a set of reusable components and style guides; it's a comprehensive strategy that includes design principles, patterns, and best practices. Such systems are built with the future in mind, ensuring that the design can evolve. By defining a clear set of design standards and components, these systems provide a consistent experience across the website while allowing for flexibility and innovation.

The following are the key elements of an adaptable design system:

- **UI component library**: A collection of reusable UI elements ensures consistency and speeds up the development process

- **Style guide**: Detailed documentation on typography, color schemes, and visual elements helps maintain aesthetic coherence

- **Design principles**: Foundational philosophies that guide design decisions, ensuring that the website remains user-centric and aligned with business goals

Now, let's break down modular architecture.

Exploring modular architecture

Modular architecture complements adaptable design systems by treating each website element as an independent module that can be added, removed, or updated without affecting the whole. This approach allows for easier updates and scalability as new functionalities can be integrated seamlessly into the existing design.

Here are some of the characteristics of modular architecture:

- **Scalability**: Modules can be easily scaled up or down based on user feedback and analytics, making it easier to adapt to changing user needs
- **Reusability**: Modules designed for one project can be reused in another, saving time and resources while ensuring consistency
- **Flexibility**: Changes to one module do not necessitate changes to others, allowing for quick adaptations and iterative improvements

It is always important to ensure there's a level of balance between everything.

Balancing consistency with flexibility

The challenge in implementing adaptable design systems and modular architecture lies in balancing consistency with flexibility. While it's important to maintain a cohesive user experience across the website, the design must also allow for customization and adaptation to meet specific user needs and preferences. This balance ensures that the website can deliver personalized experiences without compromising on the overall brand identity.

Let's explore how future-proofing can be achieved through adaptability.

Future-proofing through adaptability

Adaptable design systems and modular architecture are key to future-proofing RWD. As new devices and technologies emerge, these systems allow designers and developers to incorporate new trends and functionalities with minimal disruption. This not only extends the lifespan of the website but also ensures that it remains competitive and relevant in the ever-evolving digital landscape.

In the quest for longevity in web design, adaptable design systems and modular architecture offer a robust foundation that can accommodate the rapid pace of digital change. By investing in these adaptable frameworks, organizations can ensure that their web presence is not only resilient and scalable but also poised for future innovation. This strategic foresight in design planning is crucial for staying ahead in the dynamic world of web development, where the only constant is change.

Let's explore how you can go about future-proofing.

Future-proofing against emerging web standards

In the realm of web development, staying abreast of and future-proofing against emerging web standards is crucial for ensuring the longevity and relevance of RWD. As technology evolves and new standards emerge, web designs must adapt to maintain functionality, accessibility, and user experience. This adaptive approach not only guarantees compatibility with the latest web advancements but also secures a seamless experience for users across different browsers and devices.

Web standards are guidelines and technologies are established by standard-setting bodies, such as the **World Wide Web Consortium (W3C)**, to ensure the long-term growth of the web. These standards cover a wide range of aspects, from HTML and CSS to accessibility and internationalization, providing a common language that enables web content to be consistently interpreted and displayed across different platforms.

For RWD, adherence to web standards is particularly significant. It ensures that designs are flexible and can adjust smoothly to various screen sizes and resolutions. By following established standards, developers create more accessible, efficient, and forward-compatible websites. This adherence reduces the need for costly and time-consuming redesigns as new devices and browsing environments emerge. Let's break down some future-proofing strategies:

- **Stay informed**: Regularly update your knowledge of industry trends and standards. Participate in web development communities, attend conferences, and follow the work of standards bodies such as the W3C.

- **Semantic HTML**: Use HTML according to its intended purpose. Semantic HTML not only improves accessibility and search engine rankings but also ensures your content is adaptable to future web standards.

- **Progressive enhancement**: Design your website with a base level of user experience that all browsers can access, then enhance it for more advanced browsers with additional features. This strategy ensures your site remains usable regardless of the user's technology.

- **Responsive design best practices**: Employ fluid grids, flexible images, and media queries to create designs that naturally adapt to any screen size, future-proofing your website against the continuous release of new devices.

- **Accessibility first**: Designing for accessibility from the start ensures your website can easily adapt to new accessibility guidelines and standards, broadening your site's reach and usability.

- **Test and validate**: Regularly test your website against current web standards using validation tools. This practice helps identify and rectify potential issues before they impact your users.

As new technologies such as CSS Grid, WebAssembly, and HTTP/3 evolve, integrating them into your RWD process can offer enhanced performance and new functionalities. However, it's crucial to implement these technologies in ways that don't alienate users of older systems. Offering fallbacks or alternative solutions ensures that your website remains inclusive and accessible.

Future-proofing against emerging web standards is a dynamic process that requires ongoing attention and adaptation. By embracing best practices in semantic markup, progressive enhancement, and RWD, developers can create web experiences that stand the test of time. Investing in these strategies not only enhances the immediate user experience but also safeguards your website's relevance and functionality, ensuring it remains adaptable in the face of future web developments.

Now, let's explore why Agile could be the best approach.

Staying agile in a rapidly evolving digital space

In the context of future-proofing RWD strategies, embracing a culture of continuous learning is not just beneficial; it's essential. The digital landscape evolves at an unprecedented pace, with new technologies, user expectations, and web standards emerging regularly. To stay agile and responsive to these changes, organizations and individuals alike must cultivate an environment that prioritizes ongoing education and adaptation.

The rapid evolution of digital technologies means that what's cutting-edge today may become obsolete tomorrow. Continuous learning ensures that designers, developers, and stakeholders remain updated with the latest trends, tools, and methodologies. This constant evolution of knowledge and skills is critical for devising responsive designs that meet current and future user needs effectively.

Here's how you can build a learning environment:

- **Encourage exploration and experimentation**: Foster a workspace where team members feel safe to explore new ideas, experiment with emerging technologies, and learn from both successes and failures. This environment stimulates creativity and innovation, driving the development of forward-thinking web solutions.

- **Invest in professional development**: Allocate resources for training programs, workshops, and conferences that can broaden your team's expertise. Access to such learning opportunities keeps the team informed about the latest industry trends and best practices.

- **Promote knowledge sharing**: Encourage team members to share insights, experiences, and learnings. Whether through formal presentations, casual lunch-and-learns, or internal newsletters, knowledge sharing can exponentially increase the collective intelligence of your team.

- **Leverage online resources**: Utilize the wealth of online courses, tutorials, and communities focused on web development and design. These resources can provide up-to-date information on almost any topic, from basic programming to advanced responsive design techniques.

- **Adopt Agile methodologies**: Agile development practices, with their emphasis on flexibility, collaboration, and iteration, naturally support a culture of continuous learning. These methodologies encourage teams to adapt quickly to changes, learn from each iteration, and constantly seek improvements.

Leadership plays a crucial role in cultivating a culture of continuous learning. By setting an example through their commitment to learning, leaders can inspire their teams to pursue knowledge and growth actively. Additionally, providing the necessary resources, time, and support for learning activities demonstrates a genuine investment in the team's development.

In the competitive field of web design and development, a commitment to continuous learning can be a significant differentiator. It not only prepares teams to respond adeptly to the ever-changing digital environment but also positions organizations as leaders in adopting innovative solutions that enhance user experience and engagement.

Embracing a culture of continuous learning is pivotal for staying agile in the rapidly evolving digital space. By fostering an environment that values education, experimentation, and knowledge sharing, organizations can ensure that their RWD strategies remain effective, relevant, and adaptable to future challenges and opportunities. This approach not only benefits the immediate team and projects but also contributes to the broader goal of pushing the boundaries of what's possible in web design.

Many people ask if Agile can be used within RWD. Let's explore that further.

Agile methodologies in RWD

Agile methodologies, with their emphasis on flexibility, iterative development, and collaboration, offer a powerful framework for RWD projects. In a digital landscape characterized by rapid change and technological advancement, adopting Agile approaches can help teams stay adaptable, responsive to user needs, and aligned with the latest web standards. This approach to project management and development ensures that RWD initiatives are not only current but also future-proofed against the fast-paced evolution of the digital space.

Let's explore what principles of Agile apply to RWD.

Agile principles applied to RWD

Agile methodologies prioritize working software over comprehensive documentation, customer collaboration over contract negotiation, and responding to change over following a plan. When applied to RWD, these principles emphasize the following aspects:

- **Iterative development**: Agile promotes short development cycles, enabling teams to rapidly prototype, test, and refine web designs. This iterative process is particularly beneficial for RWD as it allows designers and developers to adjust layouts, graphics, and functionalities to ensure optimal performance across devices.

- **User-centric design**: At the heart of Agile methodologies is the focus on user needs and feedback. By continuously testing designs with real users and integrating feedback into subsequent iterations, RWD projects can achieve higher usability and user satisfaction.

- **Cross-functional collaboration**: Agile encourages close collaboration between designers, developers, and stakeholders. This integrated approach ensures that all aspects of the responsive design — from aesthetics to functionality — are aligned with the project's goals and user expectations.

- **Adaptability to change**: The digital environment is subject to frequent and sometimes unpredictable changes. Agile's flexibility allows RWD projects to adapt to new technologies, design trends, and user behaviors, ensuring the end product remains relevant and effective.

Next, we'll explore how to implement Agile.

Implementing Agile in RWD projects

To successfully integrate Agile methodologies into RWD projects, teams should consider the following strategies:

- **Sprint planning**: Divide the project into short, manageable phases or "sprints," each with specific goals related to design and functionality. This structure helps teams maintain focus and momentum while allowing for adjustments based on feedback and testing.

- **Regular retrospectives**: After each sprint, conduct retrospectives to discuss what worked, what didn't, and how processes can be improved. These meetings encourage continuous learning and optimization of workflows.

- **User stories and personas**: Develop user stories and personas to guide the design process. These tools help the team maintain a user-centric approach, ensuring that the responsive design meets the actual needs of its intended audience.

- **Continuous testing and integration**: Implement continuous testing and integration practices to identify and resolve issues early in the development process. This approach is critical for RWD as it ensures compatibility and performance across a wide range of devices and screen sizes.

Next, we'll break down the benefits of Agile for RWD.

The benefits of Agile for RWD

Adopting Agile methodologies for RWD projects offers several advantages, including the following:

- **Increased flexibility**: Agile's iterative nature allows teams to incorporate emerging trends and technologies into their designs, keeping the project at the forefront of digital innovation

- **Enhanced collaboration**: The emphasis on teamwork and stakeholder involvement fosters a collaborative environment that can lead to more creative and effective design solutions

- **Improved user experience**: By prioritizing user feedback and testing, Agile methodologies help ensure that the responsive design is intuitive, engaging, and meets the users' needs

- **Efficiency and productivity**: Agile practices, such as regular standups and sprints, can improve project management, reduce waste, and increase productivity

Incorporating Agile methodologies into RWD projects equips teams with the tools and mindset to navigate the complexities of the digital world. This approach not only enhances the development process but also ensures that the end product is adaptable, user-focused, and ready to meet the challenges of tomorrow's web landscape.

Many tools and technologies can assist with flexibility. Let's take a closer look.

Leveraging tools and technologies for flexibility

In the quest to future-proof RWD strategies, leveraging the right tools and technologies plays a pivotal role in maintaining flexibility in a rapidly evolving digital landscape. The continuous emergence of new web technologies, frameworks, and design tools not only offers new opportunities for innovation but also presents challenges in keeping up with the pace of change. By carefully selecting and utilizing these tools, teams can enhance their ability to adapt, iterate, and respond to new trends and user expectations with agility.

The tools we use are incredibly important, so let's make sure we choose the right ones.

Selecting the right tools for RWD

Choosing tools and technologies that support responsive design principles is vital. The ideal tools should facilitate efficient design and development processes, offer extensive compatibility across devices, and enable seamless integration of new features or content.

The following are some key considerations:

- **Responsive design frameworks**: Frameworks such as Bootstrap or Foundation provide a solid foundation for building responsive sites with predefined grid systems and UI components that automatically adjust to different screen sizes
- **Frontend development tools**: Tools such as Sass for CSS extension and PostCSS for automating routine tasks help streamline the development workflow, making it easier to maintain and update stylesheets for responsive designs
- **Version control systems (VCSs)**: Platforms such as Git enable teams to manage changes and collaborate more effectively, especially when working on complex projects that require frequent updates and iterations
- **Content management systems (CMSs)**: A CMS with strong RWD support allows for the easy creation and management of content that adapts to various devices, enhancing the site's usability and accessibility

As is the case with technology, there is always something new to embrace.

Embracing new technologies for enhanced flexibility

To stay at the forefront of digital innovation, it's essential to explore and adopt emerging technologies that can drive the future of RWD:

- **Progressive web apps (PWAs)**: PWAs use modern web capabilities to deliver app-like experiences on the web, offering responsiveness, offline functionality, and device-specific optimizations

- **Headless CMS**: Separating the backend content management from the frontend presentation layer, a headless CMS provides greater flexibility in delivering content across multiple platforms and devices

- **API-first design**: Building web services with an API-first approach ensures that data and functionalities can be easily accessed and utilized across different frontend applications, supporting a more cohesive and integrated user experience

Now, let's break down how to streamline workflows with automation.

Streamlining workflows with automation

Automation tools can significantly reduce the time and effort required to test and optimize responsive designs across various devices and browsers. Here's a list of different types of tools:

- **Automated testing tools**: Tools such as Selenium or Puppeteer automate the process of testing web applications across different environments, helping to identify and fix responsiveness issues more efficiently

- **Build tools and task runners**: Utilizing build tools (for example, Webpack) and task runners (for example, Gulp) can automate tasks such as minification, compilation, and other preprocessing steps, streamlining the development process and ensuring code efficiency

The rapidly changing nature of web technologies necessitates a commitment to continuous learning and experimentation. Staying informed about the latest tools, libraries, and best practices enables teams to experiment with new approaches and integrate innovative solutions into their RWD projects. Online resources, community forums, and developer conferences can be invaluable for keeping up with technological advancements and gaining insights from the wider web development community.

Leveraging tools and technologies for flexibility is crucial in ensuring that RWD strategies remain effective and forward-looking. By embracing a toolkit that supports responsiveness, adopting emerging technologies, and streamlining workflows with automation, teams can enhance their agility and responsiveness to the ever-changing digital landscape. This proactive approach not only future-proofs RWD projects but also empowers designers and developers to create more dynamic, user-centered web experiences.

Summary

As we conclude this chapter, we've navigated through the essential tactics and foresight required to adapt and thrive in the evolving landscape of RWD. This chapter has armed us with the knowledge to not only predict emerging trends and technologies but also to effectively integrate them into our RWD strategies. By staying ahead of changes in user behaviors and device preferences, we've learned how to maintain the relevance and effectiveness of our digital offerings, ensuring they resonate with users today and tomorrow.

The insights into planning for the long-term evolution of RWD strategies have underscored the importance of building with flexibility and scalability in mind. Embracing new technologies — such as AI, VR, and voice interfaces — without disrupting existing user experiences highlights our ability to innovate while maintaining a solid foundation in user-centric design principles.

These lessons are invaluable for anyone looking to sustain and enhance their digital presence in a future where technological advancements continually shift the goalposts. Implementing these forward-thinking strategies ensures that our RWD efforts are not just responsive to today's needs but are also adaptable to the future's demands.

As we move forward to the next chapter, we'll transition from the technical and strategic aspects of RWD to the human element behind successful responsive strategies. This natural progression emphasizes that future-proofing our digital strategies goes beyond technology and trends.

15
Building a Responsive Culture

This final chapter explores one of the most crucial aspects underpinning successful **responsive web design (RWD)**: the organizational culture that supports it. This chapter delves into the systematic cultivation of a workplace ethos that prioritizes adaptability, continuous learning, and user-centricity. In an era where technological change is the only constant, fostering a responsive mindset becomes essential for any organization aiming to stay relevant and competitive.

Developing a responsive culture does more than just improve the quality of your web projects — it transforms the way teams think about design challenges, enhances collaboration across departments, and heightens the overall agility of your organization. Through practical examples and proven strategies, this chapter will provide you with the tools needed to build a supportive environment where a responsive approach to web design can thrive. This foundational shift not only optimizes your current projects but also prepares your team for future technological advancements.

By understanding and implementing the lessons in this chapter, you will set your team and projects on a path to sustained success, characterized by a proactive stance toward innovation and a deep commitment to delivering exceptional user experiences.

The topics covered in this chapter are as follows:

- The mindset of adaptability
- Training and continuous learning in RWD
- Cross-functional collaboration for responsive success
- Celebrating and rewarding responsiveness

The mindset of adaptability

In the dynamic field of RWD, adaptability isn't just a beneficial trait — it's a critical survival skill. As digital platforms evolve and consumer expectations shift, the ability to embrace and capitalize on change determines a team's success and longevity. This section explores why change is a constant in the tech world and how embracing this reality can transform challenges into opportunities for growth and innovation.

Change in the tech industry is driven by several factors: technological advancements, user expectations, and the competitive landscape. Let's examine these factors:

- **Technological evolution**: New devices and technologies can change the playing field overnight, making previous approaches obsolete.

- **Consumer behavior**: As user preferences evolve, so must design strategies. What worked yesterday might not resonate today.

- **Market dynamics**: Competitors might adopt new technologies that require an adaptive response to maintain market positions.

For RWD professionals, staying static means falling behind. Thus, embracing change isn't optional; it's essential.

Developing a culture that not only accepts but also drives change involves several strategic actions:

- **Continuous learning and development**: Encourage ongoing education and skills enhancement for all team members. This might include regular training sessions, workshops, or access to courses on the latest technologies and design principles.

- **Agile project management**: Implement project management methodologies that emphasize flexibility and rapid iteration, such as **Agile** or **scrum**. These frameworks help teams stay responsive to change and incorporate feedback into their workflows effectively.

- **Proactive innovation**: Schedule regular brainstorming sessions to anticipate changes and innovate solutions. These meetings should not only address current projects but also consider broader market trends and technologies on the horizon.

- **Feedback mechanisms**: Create robust channels for feedback from all stakeholders, including team members, clients, and end users. Feedback is a crucial element of adaptive strategies, as it provides direct insights into what changes are necessary and effective.

By fostering an environment that views change as an opportunity rather than a threat, organizations can reap multiple benefits:

- **Competitive advantage**: Being quick to adapt can position a company ahead of the curve, turning potential challenges into unique selling points.

- **Innovation leadership**: A proactive approach to change often leads to innovation, setting industry standards and positioning the company as a leader.

- **Enhanced team morale and retention**: Teams that are equipped to handle change confidently are more likely to feel empowered and engaged, reducing turnover and boosting morale.

In conclusion, embracing change as a constant in the world of RWD is about more than survival; it's about thriving. It enables teams to lead with confidence, innovate without fear, and navigate the digital future with agility and foresight.

Cultivating flexibility in problem-solving

Adaptability in RWD doesn't just require keeping up with technological changes — it also demands flexibility in problem-solving. This essential skill ensures that design teams can devise effective solutions in a landscape that often shifts unexpectedly. In this section, we will explore the significance of cultivating flexibility in problem-solving approaches within RWD teams and the methods to foster this indispensable skill.

In the fast-paced realm of digital design, challenges can emerge in various forms — from new user interaction patterns to updates in accessibility standards or sudden shifts in device usage statistics. Flexible problem-solving enables teams to do the following:

- Respond rapidly to unforeseen challenges without being bound by traditional methods

- Innovate more effectively, using creativity to bypass obstacles

- Adapt solutions to the evolving technological landscape and user expectations

In the next section, we will learn about how you can cultivate flexibility.

How to cultivate flexibility

Here is a list of various approaches to cultivate flexibility:

- **Encourage diverse thinking**: Promote a team culture that values diverse perspectives and solutions. Encourage team members from different backgrounds or with different skill sets to contribute to the problem-solving process. This diversity can introduce a wide range of ideas and prevent echo chamber effects where only similar solutions are considered.

- **Implement iterative design processes**: Use iterative design methods that involve multiple stages of testing and feedback. This approach allows the team to explore several solutions and refine them over time, rather than committing to a single solution from the start.

- **Foster a safe environment for experimentation**: Create an environment where failure is seen as a step toward learning and innovation. Allow team members to experiment with unconventional ideas without fear of negative repercussions if these ideas do not succeed initially.

- **Train in adaptive techniques**: Provide training sessions focused on developing skills in adaptive and critical thinking. Workshops on lateral thinking, for instance, can enhance the ability to view problems from various angles and come up with multiple solutions.

- **Utilize flexible project management tools**: Incorporate project management tools and strategies that allow for flexibility in task handling and deadline adjustments. Tools that support Agile methodologies can be particularly useful in this regard.

In the next section, we will learn about the benefits of flexibility.

The benefits of flexibility

By developing a flexible approach to problem-solving, teams can enjoy several benefits:

- **Enhanced creativity and innovation**: Flexibility allows for more creative solutions that can lead to breakthrough innovations in project work.

- **Increased resilience**: Teams that are capable of adjusting quickly to challenges are more resilient and less likely to be derailed by unexpected problems.

- **Greater satisfaction and engagement**: Team members feel more engaged and satisfied when they are part of a dynamic and responsive working environment.

In conclusion, cultivating flexibility in problem-solving is not merely about developing the ability to adapt to the changing tools and technologies in RWD. It's also about embracing a mindset that welcomes challenges as opportunities for growth and creativity. This mindset can significantly empower a team, making it not just versatile in its technical skills but also robust in its strategic approach.

Proactive versus reactive approaches

In the evolving landscape of RWD, the distinction between proactive and reactive approaches to adaptability can significantly influence a team's effectiveness and efficiency. This section delves into the comparative benefits of being proactive rather than reactive and provides strategies for teams to anticipate changes and act decisively ahead of time.

Understanding proactive and reactive approaches

In a proactive approach, teams anticipate future trends and challenges based on thorough research and foresight. They implement strategies and updates before issues manifest, thus avoiding potential pitfalls and setting the pace within their industry.

In a reactive approach, on the other hand, teams respond to changes and challenges as they arise. While this method can be beneficial in scenarios that require immediate solutions, it often means playing catch-up, which can hinder innovation and efficiency.

Now that we understand each approach, let us dive deeper into proactive approaches. In most cases, a reactive approach is taken, which is why it is important to outline why proactive approaches are more beneficial.

Advantages of proactivity in RWD

Here are the advantages of proactivity in RWD:

- **Anticipating user needs**: Proactive strategies involve detailed analysis of user behavior and emerging device trends. By understanding potential shifts in user expectations or technology, teams can develop solutions that meet these needs before they become standard demands, significantly enhancing user satisfaction.

- **Reducing crisis management**: By addressing potential problems before they escalate, proactive teams reduce the time and resources spent on crisis management. This not only improves operational efficiency but also helps maintain a calm and controlled project environment.

- **Fostering innovation leadership**: Teams that are ahead of the curve in adopting new technologies or strategies often set trends, establishing themselves as leaders in the RWD space. This leadership position can attract higher-profile projects and clients.

Those are some advantages of a proactive approach. Now, we will dive into cultivating a proactive mindset.

Cultivating a proactive mindset

Ensuring a proactive mindset is imperative. Here are some ways in which it can be achieved:

- **Continuous learning and trend monitoring**: Encourage regular training and learning opportunities focused on the latest RWD techniques and emerging technologies. Staying informed through webinars, workshops, and industry reports can equip team members with the knowledge to anticipate and innovate.

- **Strategic planning sessions**: Hold frequent planning sessions that focus not only on current projects but also on future landscape scenarios. These sessions can help teams visualize potential changes in the digital environment and strategize accordingly.

- **Feedback loops**: Implement structured feedback loops with clients, users, and team members to gather insights about the effectiveness of current RWD strategies and potential areas for improvement before they become critical issues.

- **Risk assessment and management**: Develop a robust framework for risk assessment that includes predictive analytics to foresee possible challenges in projects. Proactive risk management means preparing contingency plans and having solutions ready to deploy as needed.

While reactive approaches are sometimes necessary, especially in fast-paced or unpredictable situations, fostering a proactive culture within an RWD team can provide substantial long-term benefits. This includes not just staying competitive but also being a trendsetter in the industry, offering innovative solutions that pre-emptively solve user problems and improve digital experiences.

In essence, shifting from a reactive to a proactive mindset involves changing how the team views challenges and opportunities. It requires a broad view that considers future possibilities and prepares for them today, ultimately leading to a more resilient and forward-thinking team.

Now, we need to explore how to create an environment for constant growth.

Training and continuous learning in RWD

Building a responsive culture within an organization isn't just about adopting new technologies or methodologies; it's also about ingraining a continuous learning ethic that supports and enhances the implementation of RWD.

This section explores the essential steps for establishing a robust learning framework that enables team members to stay at the forefront of RWD practices, ensuring adaptability and competence in a fast-evolving digital landscape.

Why a learning framework matters

In the dynamic field of web design, where technologies and user expectations are constantly shifting, continuous education forms the backbone of a proactive and innovative team. A well-structured learning framework helps in accomplishing the following:

- **Keeping up with industry changes**: Rapid changes in technology and design philosophies can quickly render skills obsolete. Regular updates and learning opportunities ensure that skills remain relevant.

- **Enhancing problem-solving skills**: Exposure to a variety of scenarios within training modules improves the problem-solving capabilities of team members, enabling them to handle more complex projects.

- **Fostering innovation**: Learning new techniques and tools can inspire team members to think creatively and bring innovative ideas to their projects.

Now that we know why a learning framework matters, it is time to explore its components.

Components of an effective learning framework

A learning framework is made up of multiple components:

- **Tailored learning paths**: Each team member has unique needs and areas for improvement. Personalized learning paths allow for targeted education that is most beneficial to individual growth and the organization's goals.

- **A blend of formal and informal learning**: It's a good idea to combine structured courses and certification programs with informal learning sessions such as workshops, webinars, and peer-to-peer coaching. This variety can cater to different learning styles and enhance knowledge retention.

- **Regular skill assessments**: This involves conducting periodic assessments to gauge the effectiveness of the learning initiatives and to identify areas needing additional focus. This helps in continuously updating the learning paths to be aligned with both individual growth and the strategic aims of the organization.

- **Integration with daily work**: Learning should be closely tied to team members' daily tasks. Real-world applications of skills learned help solidify knowledge and demonstrate the practical benefits of their educational pursuits.

- **Encouragement of collaboration and sharing**: Fostering an environment where team members are encouraged to share knowledge and insights is recommended. This not only strengthens the team's overall skill set but also builds a supportive workplace culture.

In the next section, we will learn how to implement this learning framework.

Implementing the framework

Now we can explore how to implement this framework:

- **Leadership support**: Success starts from the top. Ensure that the organization's leaders are committed to fostering a learning environment. Their support will be crucial in legitimizing the program and motivating team members to participate actively.

- **Resources and tools**: Provide access to necessary learning materials and resources. This could include subscriptions to online courses, in-house training programs, external workshops, and seminars relevant to RWD.

- **Scheduling and time allocation**: Allocate specific times during work hours for learning activities. This shows the organization's commitment to training and gives team members the time they need to develop without impacting project schedules adversely.

- **Feedback and iteration**: Use feedback from team members to refine and improve the learning initiatives. Understanding what works and what doesn't allows the organization to better serve the team's learning needs and to adjust the strategies accordingly.

Establishing a comprehensive learning framework for RWD is not just an investment in individual team members, but a foundational strategy for building a resilient and innovative organization. By continuously fostering skill growth and adaptability, companies can better prepare for the demands of the future digital landscape, ensuring that their teams not only keep up with the pace of change but lead the way in innovation and user satisfaction.

Integrating RWD skills into professional development programs

Integrating RWD skills into professional development programs is a strategic approach that not only enhances individual competencies but also bolsters the organization's capability to meet modern web standards and user expectations.

This section delves into the benefits and methodologies of weaving RWD skills into the fabric of professional development, ensuring that the workforce is not only proficient in current technologies but also primed for future advancements.

Why integrate RWD Skills?

RWD remains at the forefront of creating efficient, user-friendly online environments that cater to a diverse range of devices. By integrating RWD skills into professional development, companies can achieve the following:

- **Increased competitiveness**: Employees skilled in RWD contribute to building more adaptable and robust websites, which can lead to a competitive advantage in the digital marketplace.

- **Enhanced team versatility**: When team members are well-versed in RWD principles, they are better equipped to collaborate on various projects, reducing dependency on specialized roles and fostering a more flexible workforce.

- **Meeting evolving user demands**: As users increasingly access content across multiple devices, having a team that is proficient in RWD ensures that user engagement remains high across all platforms, enhancing user satisfaction and retention.

That concludes the list of reasons why we should integrate RWD.

Key strategies for integration

Here is a list of some key strategies to adopt integration:

- **Curriculum development**: Develop a comprehensive curriculum that covers essential RWD skills, from basic principles to advanced techniques. Ensure it includes hands-on projects that reflect real-world challenges to facilitate practical learning.

- **Professional training workshops**: Organize workshops led by experts in RWD. These sessions should provide detailed insights into the latest trends and tools in responsive design and offer a platform for employees to interact and learn from industry leaders.

- **Cross-disciplinary learning**: Encourage learning across different departments. For example, involve designers, developers, and project managers in joint training sessions. This promotes a holistic understanding of how various roles contribute to and benefit from effective RWD.

- **Use of modern learning tools**: Employ modern learning tools such as virtual classrooms, online tutorials, and interactive simulators that provide employees with the flexibility to learn at their own pace and according to their individual learning styles.

- **Regular updates and refresher courses**: Technology and best practices in the realm of RWD are constantly evolving. Offering regular updates and refresher courses can help keep the workforce informed about the latest developments and ensure that their skills remain relevant.

That is a wrap on the list of key strategies.

Practical steps to implement

This is a list of practical steps that can be taken to help with implementation:

1. **Assess skill gaps**: Conduct assessments to identify current skill gaps within the team. This will help you tailor the development programs to be most effective in elevating the team's RWD capabilities.

2. **Set clear objectives**: Define clear learning objectives and outcomes for RWD training. This helps in measuring the effectiveness of the training programs and ensures that they meet the desired goals of enhancing the skills that are relevant to the organization's needs.

3. **Provide learning resources**: Ensure that employees have access to necessary resources, such as textbooks, software, and online resources, which support self-study and continuous learning.

4. **Encourage certifications**: Motivate employees to obtain certifications in key areas of RWD. This not only validates their skills but also encourages a deeper engagement with the subject matter.

5. **Monitor progress and provide feedback**: Regularly monitor progress through quizzes, project submissions, and informal feedback sessions. Providing constructive feedback helps learners correct mistakes and refine their skills over time.

Integrating RWD skills into professional development initiatives represents a proactive approach to not only enhancing individual capabilities but also ensuring that the organization remains adaptable and forward-thinking in its digital strategies.

By committing to continuous learning and improvement in RWD, companies can better navigate the complexities of today's web environment and lay a strong foundation for future success.

Now you should know how to adopt a proactive approach to implementing a comprehensive learning framework.

Leveraging mentorship and peer learning

RWD is a rapidly evolving field that requires continuous skill enhancement and collaborative learning environments. Mentorship and **peer learning** are two pivotal methods that can significantly elevate the professional development landscape within any organization.

This section explores how these educational frameworks can be implemented effectively to promote knowledge sharing and improve RWD skills across the team.

The value of mentorship in RWD

Mentorship in RWD not only accelerates the learning process but also helps bridge the gap between theoretical knowledge and practical application. Experienced professionals can guide less experienced team members through complex responsive design challenges, providing insights that are not readily available through traditional learning channels.

Below is a list of reasons to adopt mentorship:

- **Targeted skill development**: Mentors can identify specific areas where mentees may need improvement, from foundational HTML and CSS skills to more advanced topics such as fluid grids and media queries.

- **Encouragement and support**: Having a mentor provides a safety net for junior designers and developers, encouraging them to take on challenging projects with the support of a seasoned professional.

- **Faster onboarding**: New employees can acclimate quicker to the company's workflow and standards under the guidance of a mentor, reducing the overall time to productivity.

That concludes our list of reasons to adopt mentorship.

Enhancing learning through peer collaboration

Peer learning, or **cooperative education**, involves team members working together to solve problems and share knowledge about RWD. This approach not only reinforces individual learning but also enhances team cohesion and communication.

Here is a list of different ways in which peers can collaborate to learn from one another:

- **Workshops and hackathons**: Regularly scheduled events such as workshops or hackathons can foster an environment of creativity and innovation. Peers can collaborate to tackle real-world RWD issues, enhancing their problem-solving skills and creativity.

- **Code reviews**: Encouraging peer-to-peer code reviews is another excellent way to enhance learning. Review sessions can help developers learn new coding strategies and techniques from each other, improving code quality and consistency across the team.

- **Discussion forums**: Creating an internal discussion forum where employees can post questions, share articles, and offer solutions about RWD can stimulate continuous learning and engagement.

That concludes our list of ways in which peers can collaborate.

Implementing effective mentorship and peer learning programs

Here are some ways in which you can implement effective approaches to foster collaboration in your team:

- **Structured mentorship programs**: Develop formal mentorship programs that pair less experienced designers and developers with senior personnel. Regular meetings and goals should be established to monitor progress and ensure that the mentee is gaining valuable insights.

- **Encourage informal peer interactions**: While structured programs are essential, informal interactions can also significantly contribute to the learning environment. Encourage your team to seek help from peers and to share their knowledge and experiences informally.

- **Recognize and reward contributions**: Recognizing and rewarding mentors and active participants in peer learning activities can motivate more employees to engage in these practices. Recognition can be as simple as acknowledging their efforts in team meetings or as formal as including their contributions in performance evaluations.

- **Provide resources and tools**: Ensure that all team members have access to the necessary resources and tools to facilitate effective mentoring and peer learning. This might include subscriptions to online courses, access to technical conferences, or access to the latest software tools.

- **Continuous feedback**: Gather feedback on the effectiveness of mentorship and peer learning programs regularly to make necessary adjustments. Feedback can help refine the approaches to better suit the team's needs and to cover all aspects of RWD more comprehensively.

By fostering an environment where knowledge is openly shared through mentorship and peer learning, organizations can significantly enhance the collective expertise of their teams in RWD. This not only leads to more innovative and effective design solutions but also contributes to a supportive and collaborative company culture.

Cross-functional collaboration for responsive success

Creating a seamless user experience through RWD necessitates a harmonious collaboration between designers and developers. This section explores strategies to bridge the gap between these two critical roles and ensure that they work together effectively to produce robust, scalable, and visually appealing web solutions.

Understanding the importance of cross-functional teams

In the realm of RWD, the distinction between design and development often blurs, making an integrated approach not just beneficial but also essential. Designers conceptualize the user experience and create the visual layout, while developers bring these ideas to life, ensuring functionality across various devices and platforms.

Here are two key parts of effective teamwork:

- **Shared goals**: Establishing common objectives at the outset of a project can align team members toward a unified vision. For RWD, this might mean prioritizing accessibility, maintaining brand consistency across devices, or optimizing loading times.

- **Communication**: Frequent and clear communication is vital. Regular meetings and updates can help both sides understand the challenges and constraints related to their tasks.

Techniques to enhance collaboration

There are various techniques that can help create natural collaboration:

- **Joint workshops and training sessions**: Conducting workshops that involve both designers and developers can foster mutual understanding and appreciation of each other's work. Training sessions might cover topics such as the latest CSS frameworks, design prototypes, or usability testing methods.

- **Unified toolsets**: Utilizing design software that is compatible with development tools can drastically reduce the translation gap. Tools such as **Figma**, Sketch, or Adobe XD that allow for exporting code-ready layouts, or front-end frameworks that include design kits, can bridge the technical divide.

- **Design systems**: Implementing a comprehensive design system can standardize the visual and functional elements of projects. This shared reference point helps designers and developers produce consistent outputs and reduces the scope for misinterpretation.

That concludes the list of ways to help create collaboration.

Building a collaborative culture

Now that we know various techniques for collaboration, we will explore how to introduce it at a cultural level:

- **Cross-disciplinary teams**: Organize project teams that include members from both disciplines. This arrangement can lead to early identification of potential issues and innovative solutions that might not have emerged in siloed teams.

- **Role-sharing initiatives**: Encouraging professionals to step into each other's roles temporarily can promote empathy and insight. For instance, designers could code small features, while developers might try their hand at basic design tasks.

- **Reward collaboration**: Recognize and reward teamwork and collaborative successes. Highlighting projects where cross-functional collaboration led to exceptional outcomes can motivate teams to strive for joint achievement.

- **Feedback loops**: Regular feedback sessions where designers and developers review each other's work can help catch issues early and reinforce the collaborative process.

- **Social bonding activities**: Foster a team spirit through non-work-related activities. Social events, team-building exercises, and informal gatherings can improve interpersonal relations and communication, which translate into better collaboration at work.

By effectively bridging the gap between design and development, organizations can enhance their RWD capabilities, leading to websites that are not only technically sound but also visually compelling and user-friendly. This integrated approach ensures that both the back-end functionality and the front-end experience meet the high standards expected by modern web users.

The role of project management in enhancing team collaboration

Effective project management is pivotal in fostering a collaborative environment, especially in RWD projects that require cross-functional teamwork between designers, developers, and other stakeholders.

This section discusses how strong project management practices can enhance team collaboration and lead to more successful project outcomes.

Centralizing communication and coordination

Communication is the backbone of success when working with teams. Here is a breakdown of some roles and tools that project managers rely on to increase communication:

- The project manager acts as a communication hub, ensuring that all team members are aligned with project goals and timelines. They facilitate meetings, disseminate updates, and make sure that everyone has access to the information they need when they need it.

- Using project management software such as **Asana**, **Jira**, or **Trello** can help centralize communication and keep track of progress and responsibilities. These tools are invaluable in managing tasks and deadlines, especially when handling complex projects that require tight coordination between diverse teams.

Those are only some of the ways that project managers and tools can help streamline communication.

Structuring the workflow

Structuring workflows and ensuring clarification around roles and responsibilities is crucial for success.

Clear roles and responsibilities

Defining each team member's role clearly can significantly reduce overlaps and gaps in the project workflow, which are often major sources of friction in cross-functional teams.

A well-structured workflow aids in minimizing misunderstandings and ensures that both design and development teams can work in sync toward a common objective.

Milestones and checkpoints

Setting up regular milestones can provide ongoing goals for the team and help maintain momentum. Checkpoints serve as opportunities to assess progress, address potential issues, and make adjustments to the project plan as necessary.

Fostering a collaborative spirit

Here are some ways to help foster a collaborative spirit:

- **Encourage team input**:
  - Project managers should encourage input from all team members, regardless of their role. This inclusive approach can lead to innovative solutions and increases the team's overall commitment to the project's success.
  - Regular brainstorming sessions can help in ideating and problem-solving, especially when they are conducted in an open, non-judgmental environment.
- **Build trust and accountability**. Trust is a critical component of any successful team. Project managers can build trust by being transparent about project challenges and decisions. Additionally, holding each team member accountable for their contributions reinforces a sense of responsibility and dependability among the team.

There are many more ways but those are some of my favorites.

Leveraging Agile methodologies

Just like with RWD and software development, we can leverage Agile methodologies in many other areas.

Adaptable planning

Agile project management methodologies are particularly effective in projects involving RWD due to their flexibility and focus on iterative development. Agile allows teams to adapt quickly to changes in project scope or objectives, which is common in dynamic web development environments.

Sprints and scrums can help keep the team focused and productive, while also providing regular feedback loops for continuous improvement.

Cross-functional teams

Agile encourages cross-functional team structures that combine skills from various disciplines. These teams are designed to be self-sufficient, which can enhance collaboration and speed up the decision-making process.

Those are just some other areas where Agile has been proven to be effective.

Project management as a catalyst for innovation

By effectively managing projects, project managers not only ensure that tasks are completed on time and within budget but also foster an environment where creativity and collaboration are at the forefront. This approach not only improves the quality of the final product but also enhances team satisfaction and cohesiveness. As RWD projects often require quick adaptations and frequent updates, having a robust project management strategy ensures that the team remains agile, cohesive, and ready to tackle new challenges as they come.

Creating a RWD that truly resonates with users requires more than just technical skills; it requires a fusion of diverse expertise from various fields. This section explores how cross-functional collaboration can harness this diversity to foster innovation and deliver solutions that are not only effective but also forward-thinking.

The power of cross-disciplinary teams

Combining different perspectives is a powerful approach for teams.

RWD projects benefit immensely from the integration of varied perspectives. For instance, while designers focus on user experience and aesthetics, developers concentrate on functionality and performance. When these perspectives combine, the result is a more holistic approach to web design.

Including professionals from marketing, content strategy, and user psychology can further enrich this blend, leading to a deeper understanding of user needs and behaviors.

There is also a lot of synergy that can be created through collaboration.

The magic happens when team members not only share their expertise but also synergize it to explore new possibilities. This synergy often leads to breakthrough ideas that might not emerge in a more homogenized team setting.

Effective project management encourages these interactions, ensuring that the diverse strengths of each team member are utilized.

Now, let's break down how we could structure a creative and fruitful idea session.

Structured ideation sessions

Workshops and general brainstorming sessions are incredibly powerful.

Regularly scheduled workshops or brainstorming sessions are crucial for sparking creativity. These sessions should be structured to encourage free thinking and should aim to break down the silos between different departments.

Techniques such as design thinking and mind mapping can facilitate creative problem-solving and are especially useful in mixed-discipline groups.

Feedback loops are also a great way to structure sessions. Incorporating feedback loops into the project life cycle allows for continuous improvement and helps ensure that all team members have a voice in the process. This is particularly important in cross-functional teams where people with different areas of expertise may have different priorities and insights.

Now let's explore how you can encourage effective risk-taking within your teams.

Encouraging risk-taking

Calculated risk-taking is potentially one of the most effective things that you can encourage as a leader.

A culture that encourages risk-taking and views failures as learning opportunities is essential for innovation. Leaders should emphasize the value of experimental approaches and the insights that can be gained from them, even if those approaches don't always lead to success.

Such a culture encourages team members to step outside their comfort zones and propose unique solutions that could potentially revolutionize aspects of the design.

Now, let's explore how technology can aid collaboration.

Utilizing technology to enhance collaboration

Tools such as Slack, Microsoft Teams, or Asana can help bridge the gap between different teams by providing a shared space for communication and project tracking.

Additionally, using shared digital whiteboards, or design collaboration software such as Sketch or Figma, can help visualize ideas and foster a more inclusive discussion where everyone can contribute regardless of their physical location.

We can even begin to implement AI to improve our decisions.

Integrating AI for enhanced decision-making

Leveraging AI can assist in decision-making by providing data-driven insights and predictions that can further refine design choices. AI tools can analyze user interaction data to suggest optimizations for better engagement across different platforms.

So, by embracing the diverse expertise within a team, organizations can create more comprehensive and innovative RWDs that are truly user-centric. This not only enhances the user experience but also strengthens the team's ability to function effectively in a multi-disciplinary environment. As the digital landscape continues to evolve, fostering this kind of adaptable and collaborative culture will be the key to maintaining a competitive edge and achieving long-term success.

It's time to understand how we can best celebrate and reward responsiveness.

Celebrating and rewarding responsiveness

In any team-driven environment, recognizing individual contributions is crucial for maintaining high morale and encouraging a productive workplace culture. This is especially true in the dynamic and often complex field of RWD, where collaboration and individual effort play key roles in the project's success.

The importance of individual recognition

Below is a list of reasons why individual recognition is important:

- **Motivation and engagement**: Recognition helps in motivating team members and engaging them more deeply in their work. Celebrating individual achievements can lead to increased job satisfaction and a sense of ownership over one's contributions, which is vital in a field as collaborative and detailed as RWD.

- **Enhanced productivity**: When team members feel that their efforts are acknowledged, productivity often increases. This can lead to faster completion times for projects and higher quality outputs, as team members are incentivized to maintain or improve their commendable performance.

- **Strengthened team dynamics**: Recognizing individual efforts can also strengthen team dynamics. It fosters an environment of mutual respect and collaboration, where team members feel valued and supported by their peers and leadership.

I know that there are many more reasons but those are the ones that stand out to me.

Effective recognition strategies

Here is a list of effective ways to give recognition:

- **Personalized acknowledgments**: Tailoring recognition to each individual can significantly enhance its impact. Whether this is accomplished through personalized notes, public acknowledgments in meetings, or relevant rewards, showing that you pay attention to individual needs and achievements can make recognition more meaningful.

- **Integrating recognition into regular workflows**: Making recognition a regular part of meetings or reviews helps integrate it into the team's routine. This consistent recognition can help maintain long-term motivation across your team.

- **Diverse forms of recognition**: Not every team member may value the same forms of recognition. Offering various options, from financial bonuses to additional time off or public accolades, can cater to diverse preferences and enhance the overall effectiveness of your recognition strategy.

Next, we'll learn how to celebrate achievements and the importance of doing so.

Celebrating achievements

Recognition in the workplace is not just about giving credit where it's due; it's about building a work environment that respects and appreciates the hard work of every team member. In the world of RWD, where projects can be technically challenging and deadlines can be tight, a culture that actively celebrates individual contributions can be the key to maintaining a motivated, productive, and harmoniously collaborative team.

Incentivizing team achievements in responsive outcomes

A responsive and adaptive work culture is not just built through the actions of individuals alone but also through the collective achievements of the entire team. In the context of RWD, where outcomes are highly visible and directly impact user experience, incentivizing team achievements can play a crucial role in fostering an environment of collective effort and shared success.

However, why should team achievements be incentivized? Let's look at some of the key reasons:

- **Collective responsibility**: Incentivizing team achievements reinforces the idea of collective responsibility. It encourages team members to work collaboratively, understanding that the success of the project depends on the harmonious integration of all individual contributions.

- **Boosting morale**: Group incentives contribute to a positive team spirit and can significantly boost morale. When the whole team is rewarded for the success of a project, it creates an atmosphere of mutual support and celebration that is essential for sustained team motivation.

- **Enhancing creativity and innovation**: By rewarding team achievements, you can encourage team members to think outside the box and work together to find innovative solutions to design challenges. This collaborative creativity is vital in RWD projects, which require innovative thinking to ensure that designs work seamlessly across different platforms.

Those are some of the key reasons why achievements should be incentivized.

Strategies for incentivizing team achievements

Here is a list of potential incentives for team achievements:

- **Team bonuses**: Implementing team bonuses for meeting or exceeding project goals can be a highly effective incentive. These bonuses should be structured so that they are seen as both attainable and directly linked to clear and measurable project outcomes.

- **Celebration events**: Organizing events to celebrate the successful completion of a project not only serves as a reward but also helps strengthen team bonds. These events could range from a simple lunch out to a full-day team-building retreat, depending on the scale of the achievement.

- **Recognition in company-wide communications**: Acknowledging team achievements in company-wide communications, such as newsletters or all-hands meetings, raises the profile of the team's hard work and dedication to a broader audience, enhancing the sense of achievement.

- **Professional development opportunities**: Offering opportunities for professional development as a reward for team achievements can also be a great incentive. This could include sponsoring team members to attend industry conferences and workshops, or funding advanced training courses in relevant areas.

Those are some potential ways to introduce incentives for team achievements.

Case studies and real-world examples

Here is a list of some ways to help determine where recognition is due:

- **Highlighting successful projects**: Using internal case studies or post-mortem presentations that detail what was achieved and how challenges were overcome can help in recognizing and understanding the complexities involved in a project.

- **Feedback loops**: Establishing feedback loops where team achievements are regularly reviewed and discussed can help in maintaining focus on continuous improvement and recognition.

Incentivizing team achievements is a critical component of fostering a responsive culture within RWD teams. It not only encourages a collective approach to problem-solving and creativity but also helps in maintaining high levels of team morale and motivation.

By implementing thoughtful and varied incentive strategies, leaders can ensure that their teams are not only productive but also deeply invested in the success of their projects, driving the innovation that is essential in RWD.

Creating a culture of continuous improvement

In the dynamic field of RWD, the ability to continuously improve is not just a benefit — it's a necessity. Establishing and nurturing a culture that celebrates and rewards responsiveness and continuous improvement is pivotal. This cultural foundation empowers teams to evolve and adapt their strategies to meet ever-changing technological demands and user expectations.

Let's look at why continuous improvement matters:

- **Staying competitive**: In an industry driven by rapid technological advancements, staying static means falling behind. A culture of continuous improvement keeps a team alert and ready to incorporate new trends and technologies, thereby maintaining its competitive edge.

- **Enhancing skill sets**: Continuous improvement involves constant learning. For RWD teams, this means regularly updating their knowledge base not just about new coding techniques or design trends but also about emerging user behaviors and expectations.

- **Optimizing processes**: A commitment to continuous improvement helps in refining processes over time. This could mean making workflows more efficient, improving communication channels, or enhancing collaboration techniques — all of which are crucial for the streamlined execution of RWD projects.

Those are only some of the reasons why continuous improvement matters.

Strategies to foster continuous improvement

Here are some strategies that you can use to help foster continuous improvement:

- **Regular training sessions**: Conducting regular training sessions is one of the most direct ways to foster skill development and encourage ongoing education. These sessions can be formal, such as workshops led by external experts, or informal, such as peer-led sessions that involve sharing insights from recent projects.

- **Feedback mechanisms**: Effective feedback mechanisms are crucial for continuous improvement. This includes not only feedback from clients or end users but also inter-team feedback where peers can suggest improvements in a constructive environment.

- **Reward innovations and improvements**: Recognizing and rewarding individuals or teams who come up with innovative solutions or notable improvements in processes can significantly boost motivation. Rewards can be as simple as public acknowledgment in a team meeting, or more formalized through bonus schemes and awards.

- **Incorporate improvement metrics into performance reviews**: To truly embed continuous improvement in the culture, it should be part of performance evaluations. Setting specific improvement targets and reviewing these during performance appraisals can help institutionalize the importance of ongoing development.

Next, let's explore some terms we should always keep in mind.

- **Case studies:** Illustrating successful instances of problem-solving or effective adjustments made to project strategies can inspire teams and highlight the tangible benefits of adopting a proactive improvement stance.

- **Continuous learning resources**: Providing access to ongoing learning resources, such as subscriptions to industry publications, access to training portals, or budgets for attending relevant conferences can also support continuous improvement.

Creating a culture that values and encourages continuous improvement is essential for any organization that is committed to excellence in RWD. By fostering an environment where improvement is recognized and rewarded, training is consistent, and feedback is constructive, RWD teams can remain flexible and innovative. This approach not only enhances the team's skills and processes but also ensures that the organization can adapt to new challenges and opportunities as they arise, maintaining its relevance and efficacy in the fast-evolving digital landscape.

Summary

In this chapter, we explored which tools, techniques, mindsets, and approaches can help ensure a healthy environment for learning, growing, and continuous improvement.

Throughout the insightful journey in this book, we have delved deeply into the multifaceted world of RWD, exploring its strategic importance from the ground up. Starting with the foundational concepts and strategic overviews, we ventured through practical applications, including the nuances of mobile-first strategies and the intricacies of design systems. By teaching you to integrate principles of SEO, accessibility, and performance optimization, we aimed to equip you with the comprehensive skills needed to excel in today's digital landscape. Each chapter built on the last, ensuring the development of a thorough understanding of how to implement responsive designs that are not only functional but also enhance user engagement and satisfaction.

As we close this chapter of our learning journey, we'd like to sincerely thank you for your dedication and enthusiasm throughout this book. Whether you're a seasoned tech leader, a strategic designer, or a decision-maker looking to refine your digital strategy, we hope that the insights and best practices shared here will inspire you to innovate and lead with confidence in your future projects.

Remember, the path to mastery in RWD is ongoing, and each step you take enriches your capabilities to create more inclusive, effective, and forward-thinking digital environments. Thank you for joining us on this journey, and we look forward to seeing how you bring these concepts to life in your work.

Index

‹packt›

www.packtpub.com

Subscribe to our online digital library for full access to over 7,000 books and videos, as well as industry leading tools to help you plan your personal development and advance your career. For more information, please visit our website.

Why subscribe?

- Spend less time learning and more time coding with practical eBooks and Videos from over 4,000 industry professionals
- Improve your learning with Skill Plans built especially for you
- Get a free eBook or video every month
- Fully searchable for easy access to vital information
- Copy and paste, print, and bookmark content

Did you know that Packt offers eBook versions of every book published, with PDF and ePub files available? You can upgrade to the eBook version at packtpub.com and as a print book customer, you are entitled to a discount on the eBook copy. Get in touch with us at customercare@packtpub.com for more details.

At www.packtpub.com, you can also read a collection of free technical articles, sign up for a range of free newsletters, and receive exclusive discounts and offers on Packt books and eBooks.

Other Books You May Enjoy

If you enjoyed this book, you may be interested in these other books by Packt:

Responsive Web Design with HTML5 and CSS – Fourth Edition

Ben Frain

ISBN: 978-1-80324-271-2

- Use media queries, including detection for touch/mouse and color preference
- Learn HTML semantics and author accessible markup
- Facilitate different images depending on screen size or resolution
- Write the latest color functions, mix colors, and choose the most accessible ones
- Use SVGs in designs to provide resolution-independent images
- Create and use CSS custom properties, making use of new CSS functions including 'clamp', 'min', and 'max'
- Add validation and interface elements to HTML forms
- Enhance interface elements with filters, shadows, and animations

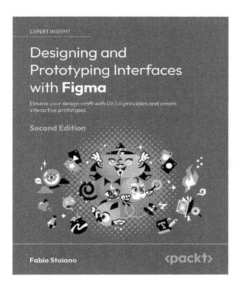

Designing and Prototyping Interfaces with Figma - Second Edition

Fabio Staiano

ISBN: 978-1-83546-460-1

- Create high-quality designs that cater to your users' needs, providing an outstanding experience
- Mastering mobile-first design and responsive design concepts
- Integrate AI capabilities into your design workflow to boost productivity and explore design innovation
- Craft immersive prototypes with conditional prototyping and variables
- Communicate effectively to technical and non-technical audiences
- Develop creative solutions for complex design challenges
- Gather and apply user feedback through interactive prototypes

Packt is searching for authors like you

If you're interested in becoming an author for Packt, please visit authors.packtpub.com and apply today. We have worked with thousands of developers and tech professionals, just like you, to help them share their insight with the global tech community. You can make a general application, apply for a specific hot topic that we are recruiting an author for, or submit your own idea.

Hi!

I am Harley Ferguson, author of *Strategic Leadership in Responsive Web Design*. I really hope you enjoyed reading this book and found it useful for increasing your productivity and efficiency.

It would really help me (and other potential readers!) if you could leave a review on Amazon sharing your thoughts on this book.

Go to the link below or scan the QR code to leave your review:

https://packt.link/r/1835080782

Your review will help us to understand what's worked well in this book, and what could be improved upon for future editions, so it really is appreciated.

Best wishes,

Harley Ferguson

Download a free PDF copy of this book

Thanks for purchasing this book!

Do you like to read on the go but are unable to carry your print books everywhere?

Is your eBook purchase not compatible with the device of your choice?

Don't worry, now with every Packt book you get a DRM-free PDF version of that book at no cost.

Read anywhere, any place, on any device. Search, copy, and paste code from your favorite technical books directly into your application.

The perks don't stop there, you can get exclusive access to discounts, newsletters, and great free content in your inbox daily

Follow these simple steps to get the benefits:

1. Scan the QR code or visit the link below

https://packt.link/free-ebook/9781835080788

2. Submit your proof of purchase

3. That's it! We'll send your free PDF and other benefits to your email directly

www.ingramcontent.com/pod-product-compliance
Lightning Source LLC
Chambersburg PA
CBHW080621060326
40690CB00021B/4772